D0885970

Improving Your
SOIL

DISCARDED

Improving Your SOIL

A Practical Guide to Soil Management for the Serious Home Gardener

KEITH REID

FIREFLY BOOKS

A FIREFLY BOOK

Published by Firefly Books Ltd. 2014
Copyright © 2014 Firefly Books Ltd.
Text Copyright © 2014 D. Keith Reid

First printing

Publisher Cataloging-in-Publication Data (U.S.)

Reid, Keith

Improving your soil : a practical guide to soil management for the serious home gardner / Keith Reid.

[272] p. : col. photos. ; cm.

Includes index.

Summary: The steps to take to achieve the perfect soil base in which to grow plants are clearly demonstrated by this author. Included are directions on amending poor soil, modifying mediocre earth, aerating compacted topsoil and substrates, and testing pH levels to enable gardeners with small and medium-sized gardens to nurture their plants and promote more abundant growth.

ISBN-13: 978-1-77085-226-6 (pbk.)

1. Garden soils. 2. Soil management. I. Title.

635.0489 dc23 S596.75.R453 2014

Library and Archives Canada Cataloguing in Publication

Reid, Keith

Improving your soil : a practical guide to soil management for the serious home gardener / Keith Reid.

Includes index.

ISBN 978-1-77085-226-6

1. Garden soils. 2. Soil management. 3. Gardening. I. Title.

S596.75.R34 2013   635'.0489   C2013-901237-0

Published in the United States by
Firefly Books (U.S.) Inc.
P.O. Box 1338, Ellicott Station
Buffalo, New York 14205

Published in Canada by
Firefly Books Ltd.
50 Staples Avenue, Unit 1
Richmond Hill, Ontario L4B 0A7

Cover, interior design and typesetting:
Gareth Lind, LINDdesign
Illustrations by Nick Craine

Printed in China

The publisher gratefully acknowledges the financial support for our publishing program by the Government of Canada through the Canada Book Fund as administered by the Department of Canadian Heritage.

Photo Credits

Cover: © Shutterstock/Konstantin Sutyagin

Chapter openers: © iStock.com/terminator1

Images on pages 25, 79, 95, 174, 219, 224, 226, 233: Keith Reid

Image on page 217: Ontario Ministry of Agriculture and Food (OMAF)

Images on pages 212, 214, 215, 221, 222, 225, 227, 228: International Plant Nutrition Institute (IPNI)

Image on page 136: © Shutterstock/SusaZoom

Image on page 172: © Shutterstock/Henrik Larsson

Image on page 179: © iStock.com/Antrey

Image on page 197: © Shutterstock/audaxi

Contents

Acknowledgments

No book is the work of a single individual, even if there is only one author listed on the cover, and this one is no exception. Many people helped shape my thinking about soil management through their questions and comments, but two of my colleagues deserve particular mention: Thank you, Adam and Anne. Thanks, as well, to our neighbors Frank and Janet, who were very good sports about being the "average gardeners," gave the first draft a read through and weren't afraid to point out where I hadn't explained something clearly.

Editing is a daunting task, but I have appreciated editor Tracy C. Read and copy editor Susan Dickinson's deft touch and gentle correction. It has made the book much more readable.

Writing is, by its nature, a solitary activity, so friends, both old and new, who took me away for a little while to reenergize and reconnect with the outside world are golden. Thanks to Dave and Marg and Wilco and Beth.

Finally, the most important thanks go to my wife, Jan, for her constant support and encouragement. Not only did she put up with my disappearing to the computer downstairs for hours on end (often when there was housework to be done), but she provided the first critique on the manuscript. She even tried to cure me of writing run-on sentences (although some things not even love can cure completely). This book is for you, dear, and now that it is done, I promise to help more around our own garden!

Preface

I'M PART OF a small minority who finds soil to be endlessly fascinating. Most people have a more prosaic interest in soil and, if they think about it at all, it is as a medium for growing plants. The most typical question about soil I hear is "What can I do to my soil to make my crops grow better?"

Soil scientists have been studying soil since the beginnings of scientific agriculture in the mid-1800's but we have barely begun to understand the intricate interplay between the air, water and minerals that make up soil. Each new discovery seems to create as many questions as answers.

This book is designed to help you manage your soil better so that you can grow a more bountiful garden by putting the right fixes in the right places. Soil does not exist in isolation. Unless it was delivered by a dump truck, the soil in your garden is a product of the parent material that was deposited thousands of years ago interacting with the cycles of rainfall, seasonal freezing and thawing and the plants that grew in soil. Over time, the plants influenced the soil, which in turn modified the environment for the plants.

My goal in writing this book is to demystify soil and offer practical methods that will allow you to grow a better garden. But I also hope this book will open your eyes to the many wonders of what goes on under your feet.

INTRODUCTION
Seasons of the Soil

Most of all, one discovers that the soil does not stay the same but, like anything alive, is always changing and telling its own story. Soil is the substance of transformation.

— Carol Williams, *Bringing a Garden to Life*

No MATTER WHERE you live, no matter what time of year it is, there is something happening underground in your garden. The ebb and flow of life in the soil as it responds to changes in temperature, moisture and plant growth is part of the mystery and beauty of the world beneath our feet.

Winter

Snow lies deep on the garden, and life has slowed to a crawl under the ground. The night crawlers have retreated to the bottom of their burrows, well below the frost line, walling themselves off from the worst of the weather. Insects have gone dormant or are spending the winter as eggs or cocoons, waiting to emerge when the weather warms up. The annual plants have died, and the perennials and winter annuals are dormant and root growth has halted.

If you think life in the soil has come to a standstill, however, you are wrong. The soil may be frozen (although probably not, if the snow cover is deep), but there is still liquid water in tiny pores and thin films. Microbes in this water are sluggish but not completely dormant. They continue to consume organic carbon in the soil and respire it as carbon dioxide. This carbon dioxide dissolves in the soil water, making it slightly acidic and helping dissolve soil minerals.

Along the pathway where you've cleared the snow (or anywhere the snow has blown off the soil surface), the environment is much different. Without the insulating blanket of snow, the temperature of the ground is much colder and more of the water in the soil freezes. If there is a long period of consistently cold temperatures, a strange thing happens. A layer of ice develops at the bottom of the layer of frozen soil. As water freezes, the amount available in liquid form is reduced, which has the same effect as drying the soil, so more water flows into this area. Since the temperature is now cold enough to freeze the water, it accumulates into a lens of almost pure ice, an inch or so thick. If the cold weather persists, multiple lenses can develop at progressively greater depths. The ice lenses push up the soil above them, which collapses back to its original level with the spring thaw. This is the "frost heave" that can push fence posts or perennial plants out of the ground.

Early Spring

The days are getting longer, and the rays of the sun carry more heat, melting the snow and warming the soil. Snow reflects much of the sun's energy, but where the soil is exposed, the dark surface soaks up the warmth. Bacteria and fungi start to grow a little faster, and the first visible sign of life are the springtails. These tiny insects become active on sunny days, even when the air temperature is below freezing. Although you may not see them against the soil, you can spot them jumping about on the melting snow, which explains how they have come to be known as snow fleas.

Soon after the temperatures start to rise above freezing, the first hardy flower bulbs poke their heads aboveground. The bulbs have been poised for action, waiting for the rising temperatures, then shooting up flower stalks using the energy stored last spring. The leaves soon follow, replenishing the food stores in the bulb so that the cycle can repeat itself again next year.

As the snow melts, some runs off the surface, but most soaks into the soil. In many areas, the soil in spring becomes nearly saturated with water. This water slowly seeps deeper into the soil, except where it is blocked by ice. You may even see water flowing out of the soil as the ice lenses melt, and a balloon of water rises to the surface like magma from a tiny volcano. Anyone foolish enough to try walking across the garden now is sure to become mired in the mud.

Sap is rising in the trees, and buds are beginning to swell. Hardy perennials and grasses are turning green, and as the top growth comes to life, so do the roots. Sugars start flowing from the leaves down through the roots and into the soil, waking the slumbering giant that is the microbial population in the soil. This includes the actinomycetes, and with every spring shower, we are greeted by the rich, earthy smell created by their growth.

Planting Time

The grip of winter fades with the longer, warmer days, as water evaporates from the surface and percolates down through the subsoil. As the soil dries out, it warms up. Yet spring is fickle, and there are always setbacks. Cool nights and the occasional late frost slow the growth of the early shoots, but the soil is a tremendous insulator. As it warms and cools in response to the cycle of day and night, the changes below ground aren't as big or as fast as the changes taking place in the weather aboveground. The soil follows its own rhythm, slow and deliberate, and sets the pace for all the living things within and above it.

Soon the soil is friable enough to be cultivated and warm enough for seeds to germinate. Earthworms emerge from their winter dormancy and vigorously till the soil in search of food. As they break down organic materials and release nutrients in their excretions, our gardens benefit from their labors.

In the perennial beds, daffodils and hellebores are at the peak of their bloom, while hostas are sending up their first tentative shoots. Mats of fungal hyphae spread through last year's fallen leaves, breaking down the cellulose and lignin and releasing nitrogen and minerals back into the soil. Some of these fungi form mycorrhizal partnerships with the plants, so those minerals go directly to the plants rather than into the soil. Chemical warfare is also being waged in the soil, as the roots of some plants release chemicals that kill the mycorrhizae. Garlic mustard is a master at this, giving it a competitive advantage over native vegetation where it has become established.

Meanwhile, in the vegetable garden, the peaceful existence of the flora and fauna in the soil is literally turned upside down. Spring cultivation mixes the compost that was spread on the surface in the fall into the soil, along with lots of oxygen. In the process, the fungi get pretty badly beaten up, their network of hyphae broken into pieces. But the introduction of fresh food and oxygen encourages the microbes, which respond accordingly with a frenzy of feeding and reproduction. The upside is a release of nutrients from the organic materials into the soil solution, where plants can use them. The downside is that this hungry mob doesn't stop with the fresh materials but begins breaking down the stable organic matter as well. Some of this activity is inevitable, but too much can hurt the soil's resilience.

Summer Solstice

The longest day of the year marks the peak of activity under our feet. The sun's rays have warmed the soil, and the moist, dark earth is teeming with life. Plants are sending out new roots at a tremendous rate, creating a rich food source for soil dwellers. Bacteria and fungi rapidly populate the zone around the roots, gobbling up the sugars and amino acids the roots release. Mycorrhizae colonize the roots, expanding the volume of soil that can contribute water and nutrients to the plants. Meanwhile, nematodes, amoebas and paramecia flock to the banquet provided by the microbes in the rhizosphere. As they feed and excrete their waste, some of the nutrients they release are reabsorbed by the roots.

The population of creatures in the soil recovers quickly from any setbacks at this time of year, which is a good thing, because there can be setbacks. Seedlings in the garden have not completely covered the soil, leaving it exposed to the effects of wind and water.

When thunderclouds blot out the sun and the first fat drops of rain begin to fall, they wallop the bare soil. Each drop is like

a tiny bomb, breaking minute particles from the granular structure at the surface and throwing them up into the air. The dry
soil quickly absorbs the moisture, but as more drops follow, the
ground is pummeled by a summer deluge. In well-cared-for soil
with lots of humus to hold its structure together, the pores at
the surface drink in the water and carry it down through the
topsoil. But it's a different picture in gardens where less care
has been taken to build humus in the soil. The pounding rain
breaks down the granules, and the small particles are carried
into the pores, where they plug them. The top of the soil rapidly
becomes saturated as the pathways deeper into the soil become
blocked, further weakening the soil structure. Water begins to
run off the surface, carrying loose particles with it. As the storm
clouds pass and the sun comes out, the surface of the soil bakes
into a solid crust that blocks seedlings trying to emerge. And the
next time it rains, this crust creates runoff even faster.

Below the crust, the microbes quickly use up the oxygen in
the soil. They then start to break down nitrate to extract the
oxygen they need for their metabolism. Rather than feeding the
plants, the nitrogen released is returned to the air as nitrogen
gas or nitrous oxide. The plants must wait until more organic
matter is broken down to replenish their supply.

Dog Days of Summer

Aboveground, the plants appear to revel in the midsummer heat,
but the livin' isn't always easy below ground. Too much water at
the end of winter has turned into too little water as days pass
with no rain. Bacteria and fungi begin to die off from lack of
moisture, as the remaining populations retreat into the small
pores and the thin water films surrounding soil particles. This
population reduction releases a flush of nitrogen, just at the
time when many plants need it for their period of maximum
growth, but it also slows the breakdown of organic matter, so
this flush is short-lived.

Different plants follow different strategies to cope with
the dwindling supply of water in summer. Some go dormant,

shutting down until the fall rains start. Some grow roots deep into the soil, pulling up water that accumulated there during the winter and spring. Others develop thickened cuticles on the leaves to reduce the rate of evaporation. All these strategies slow plant growth, so our plants may not appear as lush or vigorous as we would like. These responses, however, have been selected through aeons of adaptation to maximize the plants' chances of survival.

Even earthworms are affected when water is in short supply. As the soil dries out, the shallow-dwelling earthworm curls into a ball and covers itself with a mucous coating to protect it from drying out. Dew worms retreat deep into the soil where it is still moist and cool.

Autumnal Equinox

There is a nip in the air as the days get shorter and the heat of summer passes. Sudden rain showers moisten the earth, replenishing the water that was removed over the summer. Very little water is finding its way deep into the soil, however, since the topsoil grabs and holds the rain as it falls. Only after the topsoil reaches its water-storage capacity does water start to move down through the soil profile.

As the plants in the garden wind down their growth, the creatures under the ground perk up again. The lack of moisture that has slowed microbial growth during midsummer has eased. There is more food too. The annuals are completing their life cycle, so they are no longer exuding sugars from their roots, but the dying roots and the unharvested plant parts on the surface provide a banquet for a wide range of creatures.

Earthworms are probably more active at this time of year than at any other. You can see evidence of this wherever plant residue is left on the surface. Bean leaves are a favorite. Dew worms pull the leaves that have fallen onto the soil surface into little piles, or middens, over the mouth of their burrows. These middens serve as a food store for later in the fall and also moderate the temperature and humidity in the burrows. This gathering activity can leave large parts of the soil surface bare, which is a problem for perennial plants that need deep leaf mulch for their seeds to germinate.

Not everything shuts down in the fall. The winter annual plants are just germinating, adding some new green material to cover the ground and feed the soil organisms. In some gardens, perennials that should be entering dormancy are still growing vigorously, spurred on by the lavish nitrogen additions from gardeners who have been too generous with the fertilizer or compost over the season. They shut down eventually, when the weather gets cold enough, but this excess late growth won't help their winter survival.

Late Fall

Nights are getting frosty, and the soil at the surface cools quickly. Only a few inches down, however, the ground still retains some of the summer's warmth. Perennials are moving stores of food to their roots, ready to fuel new growth in the spring. Flower bulbs are sending out roots after lying dormant all summer. Insects and underground creatures are burrowing deeper into the soil to avoid the extremes of temperature near the surface. The ones that can't go deep are generating antifreeze to keep their body fluids from freezing.

The ground below deciduous trees is blanketed by a thick layer of fallen leaves. Pull back this mat, and you will see fungal hyphae growing vigorously where the leaves touch the soil. This, along with the mites and pill bugs chewing on the leaves, is the beginning of the process of recycling nutrients back into the soil, an efficient system that allows forests to grow tall and lush without added fertilizer. Cleaning up the leaves and burning them or carting them off to the dump disrupts nature's cycle.

The water supply in the topsoil has been replenished, so when rain falls now, the water begins to move deeper into the subsoil. This water eventually replenishes the groundwater, flowing sideways to support the base flow in the creeks and rivers next summer or percolating into aquifers that supply our wells with drinking water. The soil efficiently filters solids from the water, including bacteria that could be harmful to us. This filtering process, however, does not remove some of the anions dissolved in the water, like nitrate, and these can be carried down to the aquifers. Nitrate is a naturally occurring chemical, so a little bit of nitrate leaching is to be expected, but too many gardeners using too much fertilizer or compost can increase the concentration of nitrate in our drinking water to unsafe levels.

Winter, and the Cycle Begins Anew

The ground is frozen, and the first flakes of snow drift gently down. Most life in the soil is either dormant or moving slowly, waiting for the return of spring. The alternating freezing and thawing action as the temperature fluctuates around the freezing mark creates stresses and pressures in the soil that break down clods and make the soil more friable. In this way, the soil can heal itself from some of the management mistakes we may have made over the past summer.

Snow falls, and as the blanket of white accumulates, soil that was frozen hard when the first snow fell begins to thaw as warmth rises from below. The air trapped in freshly fallen snow makes the snow an excellent insulator; even a few inches of snow is enough to protect the soil from freezing. This creates a dim, twilight world where only a tiny fraction of the sun's light penetrates, but there is enough warmth for life to carry on. Mice and voles tunnel through this zone, feeding on the leftover grasses and seeds from last summer. Tender shoots that would otherwise be killed by frost are protected. Although fungi and bacteria in the soil are not multiplying quickly, they continue to slowly break down organic matter into humus.

For the most part, however, the soil is waiting, resting in anticipation of the renewal that comes as the days begin to grow longer.

Getting Started:
What Type of Soil
Do You Have?

We come from the earth,
we return to the earth,
and in between, we garden.
— Anonymous

1

THE SOIL ON the farm where I grew up in Bruce County, Ontario, was, to put it charitably, challenging. When the glaciers retreated about 10,000 years ago, at the end of the last ice age, they left behind a mix of clay and silt, along with a smattering of rocks and stones. After a heavy downpour, the soil behaved so much like potter's clay that kids used it to make bowls or ashtrays or knobby lumps that were supposed to be animals. If farmers worked the soil when it was wet, it would dry into bricklike clods that could not be broken down into a seedbed until the rain softened them.

Even so, when this soil was treated properly, it formed a mellow seedbed that grew very good crops. Unforgiving of mistakes, it rewarded patience and careful management. The clay that stuck to our boots when it was wet held the moisture during the hot summer months, and it also held on to nutrients. As a result, we didn't have to spend as much on fertilizer as did some of our neighbors, who enjoyed nice sandy soil that could be planted two weeks earlier.

Every soil presents unique challenges and offers unique advantages. If you know your soil, it is easier to understand how it will behave in response to your management and to the weather. You can choose to work with your soil, or you can try to impose your will upon it, despite what the soil is telling you. The results, I can predict with confidence, will be quite different.

The Gardener's Tools

Essential:

- good shovel
- trowel
- sturdy knife or narrow putty knife
- spray bottle filled with clean water
- plastic bucket
- ruler or tape measure
- notebook or clipboard with paper
- file or emery to sharpen shovel blade

Optional:

- soil sampling probe or Dutch auger
- tile probe or compaction meter
- hand lens
- camera (for documentation when submitting questions to experts)
- bags for soil or plant samples
- soil maps for local area
- moist towelettes (baby wipes) for cleaning hands

Taking Inventory

Too often, the approach to soil management in gardens is simply to start adding stuff and hope that something works. While this is sometimes successful, it's a little bit like dumping random products into the gas tank when your car isn't running right or using an injector cleaner when the real problem is a broken fuel pump. Nor will adding fertilizer fix a compacted soil or address a lack of moisture. As with a car repair, spending a bit of time and effort on soil diagnostics ensures that you are headed in the right direction.

The path you follow as you assess your soil condition depends on whether you're simply conducting a "checkup" to see how things are going or whether there are specific problems you hope to resolve. There may be overlap in the observations you make in each scenario. But depending on your goals and the results of your assessment, the next steps you take will probably vary in strategy, time and money invested.

The Basic Soil Checkup

When you go to your doctor for a checkup, the visit doesn't typically start with a series of complex lab tests, though they may come later. Instead, the examination usually begins with the doctor making a few basic observations and asking a series of questions about you and your lifestyle.

Similarly, when you take the first serious look at your soil, it makes sense to start with a series of observations and questions. These should tell you whether there are obvious issues that should be addressed before you move on to more complex or expensive assessments. If nothing jumps out in this basic checkup, you can determine which (if any) tests you should undertake.

It will take time and experience before you feel comfortable interpreting what you see during your soil checkup, but the following introduction will guide you through the basic steps. Detailed information to help you understand just what you are seeing and what it means for managing your soil is presented in the following chapters.

Step 1: Get Below the Surface

To understand your soil, you must look beyond what you can see when standing at the edge of the garden. Be prepared to get out the shovel and do some digging—though you don't need to dig too deep. Pick a series of three to five sites in different areas (depending how big the garden is and how variable), and dig down until you see changes in the soil—color, texture, density, and so on. Note the depth to the boundary between the dark topsoil and the subsoil beneath.

Step 2: Engage Your Senses

While you are digging, observe how the soil looks and how it behaves. Is it a nice dark brown color and does it break apart easily, or is it pale and cloddy? The ideal soil looks very similar to a rich chocolate cake—dark, moist and crumbly. This indicates that there is a good mix of mineral and organic matter in the soil and that you have a good balance of pore space, or the amount of space between solid soil particles and soil granules—this is known as "good tilth." A healthy soil also has a living community of its own, which you may see directly as earthworms, insects and other little critters or indirectly through the burrows and casts they leave behind. Note the depth of the topsoil layer and any obvious soil layers below that.

Look at the color of the subsoil for clues. Shades of red, yellow or light brown indicate that the soil is well drained, with abundant air to oxidize the iron-containing minerals. Layers that are white or very pale gray are a sign that minerals have been leached out. As a result, low fertility or soil acidity may be a problem. Blue or gray-green colors mean that the soil is saturated much of the time and may be too wet for many garden crops. Bright red splotches, called mottles, on a dull gray background indicate that the soil alternates between flooded and dry conditions.

As you dig, pay attention to the resistance of the soil to penetration. Does the shovel go in easily, or do you have to jump on it to get it into the ground at all? Soil that is difficult to dig can be an indication of compaction. Next, pick up a handful of each layer and feel it—does it break apart easily or is it in hard clods? Just as you had trouble getting through the soil with a shovel, plant roots have a hard time growing through compacted soil. If the roots can't thoroughly explore the soil, that limits the amount of water and nutrients the plant can reach and therefore how well the plant grows.

Your soil's behavior is also strongly influenced by its texture—the mix of sand, silt and clay. Wet a handful of soil to feel whether it is dominated by gritty sand, smooth silt or sticky clay. This exercise reveals a great deal about what challenges you are likely to face in managing your soil and how it will respond to your management efforts.

Your nose can tell you a lot about the soil as well. The earthy smell of freshly turned soil in the spring is the result of active soil biology. That aroma is present only when there is the right mix of warmth, moisture and air in the soil. What is good for the critters in the soil is generally good for the plants we grow.

Other odors can help diagnose problems in the soil. The rotten-egg smell of hydrogen sulfide means wet conditions. A sewage smell can come from rotting organic matter, often accompanied by damp conditions. The pungency of ammonia probably indicates that you have been too generous with either fertilizer or manure.

In the past, some farmers would taste a pinch of soil to determine its suitability for crops. I am told that you can detect soil

The Importance of a Good Shovel

A shovel is the tool of choice for gardeners who are serious about finding out what's happening in their soil, so it is worth investing in good quality. The false economy of a cheap shovel will always come back to haunt you.

- I prefer a round-mouth shovel with a sturdy blade about 9 inches (23 cm) wide. Narrower shovels can be easier to use, but they don't move as much soil with each bite.

- A spade with a flat, square blade is fine for finishing the sides of a hole that has already been dug, but I find it doesn't penetrate the ground as well.

- Keep your shovel sharp, whichever shape you prefer. It slices more easily through the soil and any tough roots you encounter.

- The blade of your shovel should be a good thickness—the metal on inexpensive shovels is thin and soon wears out and breaks. On the best shovels, the blade is welded to a solid shank rather than wrapped around the handle, providing greater strength and durability.

Sharpening shovel with file

- Whether to buy a straight shaft or a D-handle is strictly determined by whichever design feels most comfortable in your hands. The straight handle, with its extra length, offers more leverage, but many people find the D-handle easier to grip.

- Hardwood or fiberglass handle? Fiberglass is lighter and stronger but more expensive. It is also subject to damage from UV rays if left out in the sun for extended periods of time. Hardwood is excellent for most uses; for greater strength, be sure to look for a dense, straight grain.

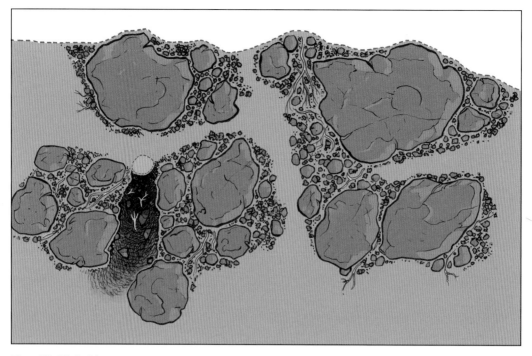

The soil is filled with pore spaces of different sizes and shapes. These dictate the way the soil will behave, and how productive your garden will be.

acidity by a sour taste and excess alkalinity by a bitter taste. Today, lab tests have pretty much replaced this method. Not only are these tests more accurate, but we are now more aware of things in the soil that we might not want to ingest.

Step 3: Listen to What Your Plants Are Telling You

If plants are growing in the area where you are digging, note the appearance of their roots. Healthy roots are white to pale cream (the color may be obscured by soil clinging to them) and grow downward at a fairly consistent angle. Look carefully to see whether the roots have changed direction or are flattened and deformed where they have grown around clods or through cracks. These are signs of poor soil structure or compaction—the soil is too hard to penetrate. Short, stubby or discolored roots can indicate problems with soil acidity, too much or too little fertility or feeding by various critters.

The vigor of your plant's top growth can provide clues about what is happening below the surface as well. Stunted growth may be a sign of insects or disease; it could also indicate a change in soil conditions that has led to a shortage of water

Healthy plant growth starts with the soil. Most roots occur in the topsoil, although some grow much deeper. Any roots that suddenly begin to grow horizontally rather than growing deeper into the soil have hit a compacted layer.

or nutrients. Deficiencies of some nutrients produce distinctive symptoms in the top growth of some species, while other nutrient deficiencies can be revealed only through a chemical analysis of the soil or the plant.

Step 4: Look at the Big Picture

Stand and stretch after digging, and as you do, take a look at the landscape around you. It can tell you a lot about what is going on in your soil. Are you at the top of a hill, where water always flows away, or in a hollow, where it accumulates? Is there a place for excess water to go, or does it have to percolate down through the soil to get away? Are there variations in growth across your garden? Are some areas producing lush, healthy plants while other areas are struggling?

Finally, consider where you live. What is the weather like? Are there definite seasons, or is it temperate year-round? Is there lots of rain, or is it usually dry? When it does rain, is it a sudden thunderstorm followed by weeks of hot sun, or do you receive continual small showers and drizzle? All these factors influence how your soil behaves and how it will respond to management.

Step 5: How Has the Soil Been Managed in the Past?

What has been grown in your garden in the past and how the soil has been treated—or mistreated—make a big difference in the challenges you face bringing the soil into good productivity. Have you purchased a property in the country and want to create a garden in an old pasture? Do you want to improve a garden that you have cultivated for many years with lots of TLC? Or are you trying to start a garden in what passes for soil in a new subdivision? If you know the history of your garden, make a note of what plants have been grown, what fertilizer or manure has been applied and what did well or poorly. These details will help reveal the potential limitations of your soil.

No matter where you are, some basic principles of soil management apply, but at some point you will be dealing with scenarios specific to your local area. My goal is to identify situations that demand local information and to point you to reliable sources for that information.

The Advanced Soil Checkup

The basic soil checkup focuses on what you can determine with tools that you probably have on hand. Many common soil problems can be diagnosed this way. Eventually, however, you may find that you need more detail than the basic checkup can provide. That's when you really need to dig in to your soil problems.

Step 1: Check for Compaction

In compacted soil, the pores have been squeezed shut—there is nowhere for water to drain and nowhere for air to enter the soil or for roots to grow. A compact soil may be fine if you're digging a house foundation, but it's not what you need to grow a garden. You can easily check for compaction using a shovel, a soil sampling probe or a compaction meter, or by simply digging a hole and testing the various layers with a screwdriver. The goal of this process is to detect layers with greater resistance to penetration. The top layer of soil may have been loosened by tillage, but it is not uncommon to find a layer below where the soil has been

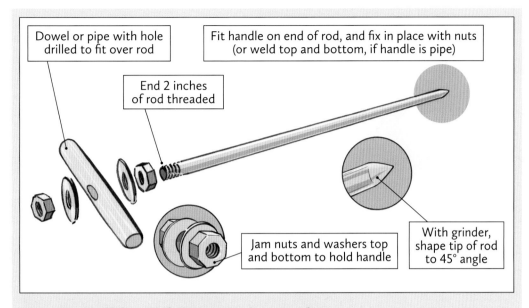

Dowel or pipe with hole
drilled to fit over rod

Fit handle on end of rod, and fix in place with nuts
(or weld top and bottom, if handle is pipe)

End 2 inches
of rod threaded

Jam nuts and washers top
and bottom to hold handle

With grinder,
shape tip of rod
to 45° angle

How to Build a Soil Compaction Probe

A commercially available soil compaction probe (available from a number of suppliers for between $200 and $300) has the advantage of consistent construction, but unless you use it very carefully, the readings on the dial can be as misleading as they are helpful. It's possible to build a tool that is just as effective at a much lower cost. All you need is a ¼-inch-diameter (6 mm) round steel rod, about 3 to 3½ feet (1 m)

in length, with the top couple of inches threaded, and a hardwood dowel or metal pipe about 12 inches (30 cm) long to serve as a handle.

Drill a hole through the middle of the dowel, and mount one end of the rod firmly in this hole—it must be attached so that you can pull the rod out of the soil as well as push it in. Grind the other end of the rod to a point with the sides at a 45-degree angle.

compacted. It may show up on the surface as an area where the water does not drain away as quickly in the spring. In that case, planting is delayed. This scenario increases the risk for further compaction, since wet soils pack more easily than dry ones.

The best time to check whether your soil is compacted is when it is moist but not wet. A dry soil can be hard to penetrate even if there is lots of pore space, so uneven soil moisture can cause misleading results. If you use a penetrometer where there has been rain on a very dry soil, it tells you how far the moisture has penetrated rather than how compact the soil is, since it slides easily through the moist soil and then stops when it reaches the dry layers.

Using either a commercial penetrometer or your homemade compaction probe, stand with your feet about shoulder width apart, with the probe in front of you so that your arms are nearly straight. Push the probe slowly down into the soil, using even pressure. You will be able to feel changes in resistance through your hands and arms; some people find it helps to close their eyes so that they can concentrate on what they're feeling. When you come to a layer where the resistance increases, stop and measure the depth. Continue pushing the probe down through the soil and measure the depth to the bottom of the compacted layer (where the resistance to your pushing suddenly decreases). Keep going to determine whether there are multiple compacted layers. Repeat this exercise in several areas to see where and how extensive the compaction is.

Step 2: Getting a Soil Sample Analyzed

You can obtain very important information about your soil from a chemical analysis of a soil sample, including the soil pH (acidity or alkalinity), electrical conductivity (salt content) and whether there is a shortage of nutrients, which can limit how well your plants grow. You can't determine these things simply from looking at or feeling the soil.

If a soil analysis is going to be useful, it must represent the

Commercially Available Analysis Kits

Many garden centers and mail-order suppliers offer kits for analyzing soil samples. While getting answers immediately rather than waiting for lab results is appealing, these kits are probably not a good option for most gardeners.

- The chemicals used to extract the available nutrients may not be suitable for the soils in your area. They work reasonably well in acid soils but not in alkaline soils.

- Getting accurate results takes a great deal of care and a level of cleanliness that is difficult to achieve when working with soil.

- Comparing subtle gradations of color in a reagent with colors on a printed chart is difficult at best and impossible if you are one of the 10 percent of males who are color-blind. To make this problem worse, organic matter dissolved from the soil adds a color that masks the color of the reagent.

In general, the cost for analysis at a properly equipped lab is money well spent.

area from which it is collected. Soil varies from place to place, so a single sample won't provide good information. This variation might be the result of the way the soil formed or of an uneven application of fertilizer or compost.

To achieve a baseline level for your garden, collect several subsamples from different parts of the garden and mix them together. The double handful of soil you send to the lab must represent the entire garden, so it is critical to collect enough subsamples—I suggest a minimum of 8 to 10.

If you are trying to discover why one part of the garden performs differently from the rest, the principle of taking several subsamples stays the same—but keep the subsamples separate.

You can collect each subsample with a shovel or spade, but it is much easier to use a soil sampling probe or an auger when you are gathering multiple samples. If you're using a shovel, remove one shovelful of soil to open a vertical face 6 to 7 inches (15–18 cm) down, then take a slice off this face with the shovel, roughly 1 inch (2.5 cm) thick. With a knife or trowel, cut a vertical piece about 1 inch (2.5 cm) wide from this slice so that you have a sample that is 1 by 1 by 7 inches (2.5 x 2.5 x 18 cm) deep. Place this sample in a clean plastic pail, fill in the hole, and move over a few paces to collect the next sample. Don't worry if the sample crumbles when you put it in the pail—you will be breaking it up to mix with the other samples anyway. Try to get the same amount of soil from each subsample to avoid skewing the results by having more soil from one area than another. The process is the same if you are using a soil sampling probe or an auger, except you simply insert the tool to the proper depth, pull it out and place the core in the pail.

Use a clean pail, as avoiding contamination of the sample is essential. Do *not*, for instance, reuse a bucket from the garden center that contained fertilizer. The pail should be made from an inert material like plastic or stainless steel, particularly if you want to check the levels of any micronutrients. A galvanized pail may be sturdy, but the zinc coating will push the zinc levels of any sample mixed in it off the scale.

Once you have collected the sample, the next step is to choose a lab that can perform the relevant analyses—and perform them

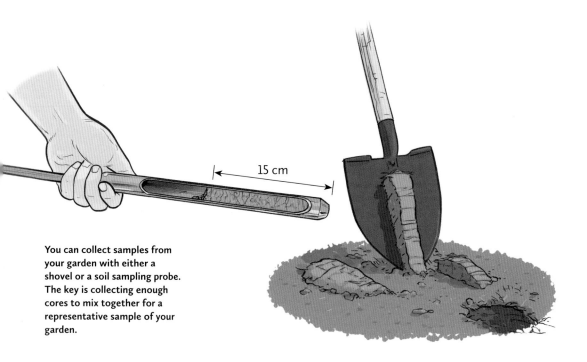

15 cm

You can collect samples from your garden with either a shovel or a soil sampling probe. The key is collecting enough cores to mix together for a representative sample of your garden.

correctly. First, be sure the lab actually tests for plant-available nutrients, which means it should be an agricultural lab, not an environmental lab.

An environmental lab is what you find if you simply look in the Yellow Pages under "soil analysis." It specializes in finding contaminants. If you request an analysis for phosphorus or potassium, for example, an environmental lab records the total amount of these elements after the sample is digested in strong acid. This is not the same as the amount a plant can use from that soil and is therefore not good information to guide your soil management. In addition, an environmental lab probably charges a significantly higher price for the analysis (in the range of $100 or more per sample) than does an agricultural lab (roughly $20 per sample). Prices vary depending on which tests are required, so check with your local lab for actual costs.

Your local department of agriculture should have a list of recommended laboratories for your area. Check there first. Visit the North American Proficiency Testing Program (www.naptprogram.org) to see whether the lab is enrolled in a quality assurance program. Appendix A provides a list of some labs you may want to check out.

Step 3: The Secrets of the Deep

It takes someone who is really enthusiastic about soil to dig a trench specifically to look at what is going on underground. However, if a backhoe is on your land for another purpose—to dig a hole for a house foundation or a water-pipe installation, for instance—a thoughtful look at the sides of the hole can be very enlightening. All the soil layers are exposed, which gives you a chance to see whether the soil texture changes deeper in the soil or whether there are changes in color that indicate where water accumulates for all or part of the year. You can also see earthworm burrows and how deeply the roots penetrate.

Realistic Expectations for Improving Soils

Throughout your life as a gardener, you must deal with the soil you have. Some characteristics can be easily changed through management. Others can be changed only with earth-moving equipment. And sometimes the only remedy is to sell your property and start over somewhere else.

In other words, there are "fixed" characteristics that determine what you can and cannot do with your soil. Other characteristics are intermediate in their response to our efforts. These intermediate characteristics can be changed over a long time period, but only after a significant investment in soil amendments and management.

Once you recognize where the challenges lie, you can better judge where to invest your effort and predict how soon you are likely to see an improvement.

"Fixed" Soil Characteristics	"Responsive" Soil Characteristics	"Intermediate" Soil Characteristics
Can't be easily changed	Can be changed through management	Can be changed, but not quickly or easily
▸ Texture	▸ Soil structure	▸ Soil organic matter
▸ Topography	▸ Drainage/irrigation	▸ Soil compaction
▸ Climate/location	▸ Fertility	▸ Soil biology
	▸ Soil acidity	▸ Water-holding capacity

Soil and Plant Diagnostics

Observing the growth of your plants and the conditions of the soil in which they are growing can provide clues about what is limiting growth or fruit production. Symptoms can vary between species, and developing the skill for this detective work can take a lifetime. Learning to see the "big picture"—what parts of the plant are affected and what types of symptoms are present—is the first step in building this skill. The tables on the following two pages will direct you to the right place to learn what is actually happening to your plants.

Timing – Planting to Emergence		
Condition of Plants	**Condition of Soil**	**Possible Causes**
No emergence	Hard surface crust; seedlings leafing out below ground	Poor soil structure, combined with heavy rain or overwatering
	Soil dry and cloddy	Poor seed/soil contact; lack of rain or irrigation
	Soil wet; may or may not have standing water	Fungal diseases causing seed to rot or seedlings to die (commonly known as "damping off")
	Firm, granular soil structure; good moisture	Poor viability of seed
Spotty emergence; some delay	Soil dry and cloddy	Poor seed/soil contact; lack of rain or irrigation
	Soil dry and powdery	Fluffy soil prevents moisture migration to seed
	Firm, granular soil structure; dry soil	• Shallow planting resulting in insufficient moisture • Salt injury from excess fertilizer or manure
	Firm, granular soil structure; good moisture	Planting too deep

Timing – Flowering and Fruiting		
Condition of Plants	**Condition of Soil**	**Possible Causes**
Few or no flowers; lush vegetative growth	Moist, rich soil	Excess nitrogen; imbalance between nitrogen and other nutrients
Ends of fruit rotting	Alternating wet and dry conditions	Calcium deficiency (in the plant, not necessarily in the soil)
Misshapen fruit; poor seed development	Any	Nutrient deficiency: potassium, micronutrients

Timing – Plant Growth

Condition of Plants	Condition of Soil	Possible Causes
Entire plant affected		
Wilting; leaves grayish and floppy	Dry soil	Drought stress
	Wet soil	Excess moisture results in oxygen deficiency to roots
Stunting	Any	Severe nutrient deficiency (nonspecific, may be phosphorus); compaction; insects feeding on roots
Yellowing	Any	Wet feet (lack of oxygen to roots); sulfur deficiency; severe nitrogen deficiency
Speckling or discolored spots on leaves	Any	Diseases (fungal, bacterial or viral)
Red or purple leaves	Any	Stress reaction of plants (anthocyanin accumulation) may indicate a phosphorus deficiency but could also be cold conditions, compaction, etc.
Leaves dropping off	Dry soil	Drought stress
	Wet soil	Lack of oxygen to roots
Older tissue affected (usually lower leaves)		
Yellowing on leaf margins		Nutrient deficiency: potassium
Yellowing on entire leaf or between veins		Nutrient deficiency: nitrogen
Speckling or discolored spots on leaves		Diseases, primarily fungal
Leaves dropping off	Dry soil	Drought stress
	Wet soil	Lack of oxygen to roots
	Any soil	Natural senescence (plant maturity)
Irregular holes in leaves; skeletonization (veins remaining); slime trails	Moist soil or heavy dew at night	Slug feeding
Younger tissue affected (usually upper leaves)		
Yellowing on leaf margins		Nutrient toxicity or salt injury
Yellowing on entire leaf or between veins		Nutrient deficiency: micronutrients
Speckling or discolored spots on leaves		Sun scald from watering in bright sun; ozone damage
Holes in leaves		Insect feeding

What Is Soil, and Why Is It Important?

When you are standing on the ground, you are really standing on the rooftop of another world.

— Dr. M. Jill Clapperton

2

It's tempting to think of soil as simply broken-down bits of rock, but while the mineral parts of soil are important, they are only a portion of the whole. To understand soil and make intelligent decisions about managing it, it helps to think of soil as a network of empty spaces surrounded by minerals and organic matter. It is in these pore spaces that everything happens.

If you were to dig up a block of soil from an old prairie or forest and measure what was in it, you would find that the pore space would comprise half its volume. If it had rained recently and the soil had just finished draining any excess water, half this pore space would be filled with water, the other half with air. Looking more closely at the solid parts of the soil, you would find that about 5 percent of its volume is made up of humus and other organic materials, largely in films either covering or mixed with the soil minerals that make up the other 45 percent.

The diagram on page 36 represents the "ideal soil" we should strive to create in our gardens but which few of us manage to achieve. More often, the pore space in our garden soil takes up far less than half the soil volume. In addition, the soil usually holds either too little or too much water when it finishes draining.

How can we get closer to this ideal? The key lies both in managing the soil so that there is more pore space and in creating the right mix of large and small pores. The small pores hold moisture much as a sponge does, so roots can easily be pulled out as needed. The large pore spaces allow excess water to drain away and air to enter the soil. Achieving this perfect mix takes a combination of tillage and biological activity, along with a healthy dose of patience.

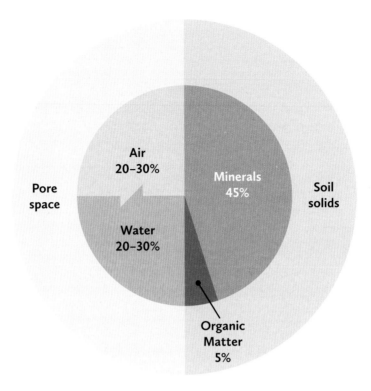

The ideal soil is composed of the correct proportion of air, water, minerals and organic matter.

What Does Soil Provide to Plants?

Support

Without the roots anchoring plants to the soil, plants would flop over on the ground instead of growing upright.

Water

The main function of plant roots is to absorb water (and nutrients) from the soil to meet the needs of the plant for transpiration. A large tree can transpire 40,000 gallons (150,000 l) of water in a year—enough to drain a small backyard swimming pool. Without the store of moisture in the soil, it would need to rain about two-tenths of an inch (5 mm) each day just to keep up with the immediate needs of the plants. Alternatively, you would have to supply that amount of water with a hose.

Nutrients

Every soil has a significant store of nutrients that plants can use. We may have to add some extra to make up for a shortfall, but this is just a small part of the total nutrient uptake by the plant. To understand this better, consider the nutrient mixtures that are needed for hydroponic production, where the growing medium (e.g., rock wool or coconut fiber) doesn't provide any nutrition. The hydroponic solution must include all the essential elements, in the proper proportion, as well as some other beneficial elements (like silicon and sodium), or the plants will not survive. In most gardens, we need to supply only a few nutrients; the rest are already there.

Air

We don't usually think of roots as having to breathe, but the cells in the roots need oxygen if they are going to perform their functions. If the soil is completely flooded, most plants will die because oxygen cannot get to the roots. Plants that are adapted to wet conditions, such as rice, have special tubes inside the stem that carry air down to the roots so that the roots don't need to get air from the outside. This group, however, does not include many of the common garden plants.

Temperature Moderation

Soil does not warm up as quickly as does the air when it is hot, nor does it cool down as quickly when the air is cold. While this can cause us to feel impatient in the spring, when the sun shines warmly and we want to get seeds in the ground (it is almost always best to wait), the "heat sink" in the soil can provide real benefits during the growing season. On cold spring nights, the warmth stored in the soil during the day can moderate the temperatures around young seedlings and prevent cold shock or frost damage. Later in the summer, the soil keeps the temperature in the crop canopy from rising too high during extreme heat, protecting the plants from stress. Again in the fall, as the weather starts to cool down, warmth released from the soil can provide protection from early frosts.

Beyond the direct benefits soil provides to growing plants, it helps us all in our day-to-day lives:

Soil Acts as a Reservoir

When it rains or when snow melts in the spring, much of the water soaks into the soil. Depending on the soil texture, this water drains through the soil either quite quickly or very slowly. Some of the water is held near the soil surface, where plants can reach it, while a portion moves deeper into the soil. The water that drains down to the water table can percolate into groundwater to replenish our aquifers or flow laterally to feed streams and rivers during the dry summer months.

Soil Acts as a Filter

Water is very effectively filtered when it percolates through the soil. Sediment and bacteria particles are caught in the small spaces between soil particles and left behind. Electrostatic charges attract and hold more potential contaminants, and the active biology in the topsoil breaks down many organic materials that could otherwise impair water quality. This is why most well water can be drunk safely just as it comes from the ground, without any further treatment.

Soil Acts as a Climate Modifier

Soil affects the climate both at the level of individual plants, as described above, and on a global scale. On the larger scale, soil is an effective carbon sink, absorbing carbon rather than releasing it into the atmosphere as carbon dioxide, a greenhouse gas. Increased levels of carbon dioxide in the atmosphere have been linked to changes in average temperatures and weather patterns. Good soil management can trap a lot of carbon by increasing soil organic matter.

Soil Creates Habitat

The soil is a diverse ecosystem on its own, but it is also home for a number of creatures that we see only during the aboveground phase of their life cycle. The cicadas that buzz from the trees on hot summer days spend most of their lives underground, emerging only to mate and lay eggs. Most crickets, grasshoppers and locusts spend part of their lives in the soil, although for many, the soil serves primarily as protection for the eggs; the nymphs emerge from the soil soon after they hatch. Mammals also use the soil as a home. Woodchucks, badgers and prairie dogs dig extensive underground burrows, coming out only to feed. These burrows, in turn, may eventually become dens for foxes or rabbits. The insulating value of the soil keeps these creatures cool in the summer and warm in the winter.

Different Soils, Different Challenges

The soil is considered as different from bedrock. The latter becomes soil under the influence of a series of soil-formation factors.
— N. A. Krasil'nikov, *Soil Microorganisms and Higher Plants*

3

As you leaf through the pages of various garden books, you will inevitably find a short section of generic soil-management information that might lead you to believe that all soils are the same. Nothing could be further from the truth. In fact, the diversity of soils mirrors the diversity of the plants that naturally grow in them. We can thank Russian soil scientist Vasily Dokuchaev for the concept of soil classification. Dokuchaev first noted in the 1880s that the soil under grasslands was very different from the soil that occurs under a pine forest or a wetland area. He determined that rather than simply being broken-down rock, soil has formed over time in response to the parent material, vegetation, climate and topography.

The implications of the differences in soil and how each soil will respond to management are myriad, but they are often ignored. More to the point, the advice given to someone who lives in an area with a strongly weathered sandy soil will not likely work for someone who lives on a clay prairie soil, and vice versa.

In this chapter, you'll find tips on how to recognize when a particular piece of advice doesn't apply to your situation and can therefore safely be ignored.

Soil-Forming Factors and their Impact on the Soil

Parent Material

As the name suggests, this is the stuff from which your soil formed. Many textbooks show diagrams of bedrock being slowly broken down through a combination of weather and the acids released by lichens and mosses until, gradually, a soil is formed. While this can happen, as with the volcanic soils in Hawaii, it is not common. Far more likely, your soil developed in sediment that was laid down by wind, water or ice.

What was left behind after the glaciers retreated or the land emerged from under water was not soil. It did not begin to form a soil until after the first plants started to grow in it. These plants added organic matter to the mineral matrix, but the nature of this matrix had a tremendous influence on the type of soil that eventually formed.

The texture of the parent material dictates how easily water drains through the soil, which, in turn, determines how quickly minerals leach out of the surface layers and into the subsoil. Sandy or gravelly parent materials, like those laid down by flowing water or in the beach ridges left along the shorelines of prehistoric lakes, allow water to drain very quickly. Under these circumstances, minerals tend to leach out, leaving soils that have low natural fertility and may be acidic.

Clayey parent materials, formed from sediments laid down in deep water, are slow to drain unless there is a good soil structure in which large cracks and pores have formed. These clayish soils retain nutrients very well, but they can also tie up a lot of added nutrients in forms not easily available to plants.

Soils that formed on glacial till, remnants from the last ice age, are a mixture of soil textures and can exhibit the best—and worst—characteristics of each. Because the ice sheets acted like big bulldozers, scooping up everything in their path, these soils tend to have lots of stones mixed in with the other materials.

If you are curious about the landforms in your area and the types of parent materials that form the basis of the local soils, check the website of your state or provincial geological survey. Most have excellent maps of the surficial geology or physiography available online or as hard copy.

The chemistry of parent materials can vary widely as well. Some started out with a rich supply of plant nutrients, while others were deficient in one or more, and that deficit influenced not only the type of plants that would grow but also their vigor. There is also a huge range of soil pH (acidity or alkalinity). Where I grew up, the ice sheets had traveled over the dolomite and limestone bedrock of the Niagara Escarpment, so there is lots of limestone mixed in with the soils. As a result, these soils are quite alkaline. In other areas, where the ice sheets had ground up granite or quartz, the deposits left behind are quite acidic.

Adding complexity to this picture is the fact that not all parent materials are composed of the same materials for the entire depth of the soil. There may be a silt loam cap over a clayey till or a gravelly outwash deposit. These soils behave very differently from each other and from a soil that is a deep silt loam.

Topography or Elevation

Closely related to the parent material is the topography of the landforms created when the parent material was laid down. Topography dictates where the water will go, how much will accumulate, how much will percolate through the soil profile—a term soil scientists use to describe the pattern of layers, or *horizons*, in the soil—and how much will run off.

Upland areas tend to be well drained, forming deep soils. Side slopes, particularly where water collects into channels, usually have shallower soils, because erosion carries away the soil as quickly as it forms. Deep soils can form at the foot of these slopes, courtesy of the eroded soil that has made its way down from higher up the slope. Depressions that remain flooded for part of the year can accumulate high levels of organic matter, because the lack of oxygen slows down the decay of plant materials. Unless the soil is drained to remove the excess water, these areas will suit only a small number of plant species that are adapted to wet conditions.

Climate

Temperature and rainfall influence the speed at which soil-forming processes take place as well as the processes that will dominate. As a general rule, all chemical reactions occur faster at high temperatures. Minerals break down faster and organic matter decomposes more quickly in hot areas than in cold ones. On the other hand, plants grow more slowly in cool areas, so there is a smaller supply of organic material to feed the decomposition. The balance, however, favors a net increase in organic matter in the soil as you move farther north or to higher elevations.

Precipitation has an even bigger impact on soil formation, particularly the balance between rainfall and evapotranspiration, which is the sum of evaporation and plant transpiration from the surface to the atmosphere. If there is more rainfall than evapotranspiration, the excess water either runs off the surface or drains through the soil. This water carries away excess salts and leaches minerals and organic materials down through the soil profile.

Where evapotranspiration dominates over precipitation, water is actually pulled up through the profile as it evaporates at the surface. This leads to the accumulation of minerals in the top layers of the soil. You can see this around ponds or sloughs in the Great Plains, where a steady supply of water percolating through the soil and dissolving minerals can lead to a crust of salt or gypsum around the shoreline.

As conditions get drier yet, the lack of water can limit or prevent plant growth. Now we're entering the realm of the desert soils, which are productive only when irrigation water is supplied. These soils are very poorly developed due to a lack of organic matter and a lack of moisture to move minerals through the soil profile.

Vegetation

While it may be obvious that the vegetation growing in a particular area is linked to the type of parent material, the topography and the climate, the vegetation also influences the type of soil that develops. Over time, the vegetation can change the soil environment enough that different plants become established, so there can be a "soil succession" in much the same way there is succession in plant communities, like a forest progressing from fast-growing poplars to the climax forest of maples and beeches.

Different types of vegetation generate different amounts of organic matter and leave it in different places in the soil profile. Where the vegetation is dominated by forests, the majority of the organic matter is returned to the soil as falling leaves or needles that decay on the surface. Prairies, on the other hand, where the vegetation is dominated by grasses, return the majority of the organic matter below the surface as decaying roots. This difference in vegetation creates markedly different soil profiles. The prairie soils have deep topsoil rich in organic matter, but the layers, or horizons, below may be rather poorly developed. Forest soils, in contrast, have relatively shallow topsoil layers, although they often start out with a layer of natural mulch at the surface. The horizons below this can be quite distinct and extend deeper into the soil. These soils tend to have less native fertility than the prairie soils and need more added nutrients to be productive.

Time

The processes that create soil operate very slowly, so the differences in time that affect soils are on the geologic scale. Pedologists (scientists who study the formation and distribution of different soil types) who travel to the James Bay Lowlands in Ontario and Quebec can study a progression of soils that have formed over the past few hundred years as the shoreline of James Bay slowly retreats. Other areas with new soils are on lava flows in places like Hawaii or in mountainous areas where glaciers are retreating.

These new soils are not typical of most areas, although there may be a lot of parallels between a suburban lawn that has been plunked directly on top of the disturbed fill around a new house and the first tentative growth on the newly emerged till at the toe of a glacier. Many of us, however, are dealing with soils that have had much longer to develop. In most of Canada and the northern United States, soils began forming after the last ice age ended, about 12,000 years ago. Farther south, along the coastal plain in the southeastern United States, it has been hundreds of thousands of years since the land was covered with ice, but even so, the climate and vegetation changed significantly during the ice ages, thereby influencing the types of soils that formed.

In general, as soils get older, the depth of the soil horizons increases, the number of horizons increases and the horizons become more distinct. Clay minerals are weathered and the minerals are leached out, changing from nutrient-rich illite, vermiculite and montmorillonite clays to less fertile kaolinite and, finally, to iron and aluminum oxides. The conditions you start out with and the ability of your soil to hold on to nutrients and water will vary greatly with the age of your soil.

Soil formation is also a balance between soil-building processes and losses through erosion. Theoretically, the soil should eventually reach equilibrium, where new soil is being formed at exactly the same rate as soil loss is occurring. In practice, however, because of shifts in climate patterns or human impacts, the soil never reaches a true state of equilibrium but is constantly changing, albeit slowly.

Your Soil Profile—the Outcome of the Soil-Forming Factors

The end result of thousands of years of soil formation is a *soil profile*, the term used to describe the pattern of layers, or horizons, in the soil.

Each soil profile is unique, but all soil profiles share some common elements. Soil scientists have developed a type of short-hand to describe these profiles, using letters to designate the various layers. While you can manage your soil just fine without knowing this system, it can be helpful for understanding the legend on a soil map when you research the soils in your area. Nearest the surface is the "A" horizon, the layer we normally refer to as topsoil. At the bottom of the soil profile, the unweathered parent material is referred to as the "C" horizon. In between are one or more "B" horizons with varying characteristics and one

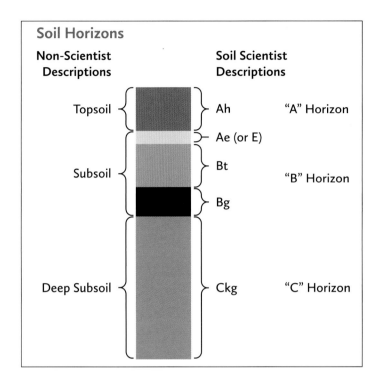

Soil scientists may use different terms for the layers in the soil than gardeners. This diagram should help make the link between the two.

or sometimes two "A" horizons. These horizons form as organic materials accumulate at the surface and water percolates through the profile carrying minerals, clay and organic matter to accumulate deeper in the soil. The differences in the B horizon have a huge influence on how the soil behaves and can be indicators of management challenges you will face.

One of the best examples of this is the Bt horizon, where clay that has washed out of the surface soil has accumulated in a distinct layer deeper in the profile. This horizon is not readily visible to the casual observer, since it is usually the same color as the layers above and below, but because of the increase in clay content, it holds more water and thus often slows drainage.

This can be a good thing in coarse-textured soils, where the extra moisture can mean the difference between good crops and drought-stressed crops.

On finer-textured soils, however, this extra moisture can slow the warming of the soil in the spring and make it more susceptible to compaction. The exception to this is soil where the extra clay in the Bt horizon creates a stronger soil structure, allowing

Decoding Soil Horizons

Soil horizon descriptions include a prefix—A, B, C, O (organic)—and usually one or more suffixes. These suffixes describe what is happening in that horizon and can be very helpful in understanding your soil.

Canadian Suffix	USDA Suffix	
h		Accumulation of organic matter
e	E	Removal of clay, organic matter, iron or aluminum
p		Plow layer or other disturbance
f	s	Accumulation of iron or aluminum oxides
g		Mottling and gleying (dull colors from lack of oxygen) due to water accumulation
m	w	Slight color or structural changes from parent material
y	jj	Cryoturbation (periodic freezing)
c	m	Strongly cemented horizon
n		Sodium salt accumulation
s		Soluble salt accumulation
k		Presence of calcium or magnesium carbonates
ss		Presence of slickensides (smooth clay coating caused by stress in high clay soils)
t		Accumulation of clay
ca	k	Accumulation of calcium or magnesium carbonates
z	f	Permanently frozen soil

channels to form through which excess water can drain.

A more extreme example of this process is the formation of cemented horizons, as with a fragipan—a horizon that often resists root penetration as well as water drainage, effectively limiting the depth of the soil. This horizon may have been cemented by iron compounds, calcium carbonates or silicate minerals, depending on the soil-forming conditions that are in place. Even if you mechanically disrupt a fragipan, it is likely to re-form within a season or two. If you suspect you are dealing with a fragipan soil, check with your local department of agriculture or extension service for advice on managing this soil effectively.

The blue-gray or gray-green color in a Bg or Cg horizon is an indicator that the soil is poorly drained and is saturated for part or all of the year. This coloration comes from the response of iron to a lack of oxygen. Iron switches from its oxidized

The Mystery of Mottles

While the dominant colors in poorly drained soils are blue or green shades and dull gray, spots, streaks and blotches of bright red or orange are often present. These are known as mottles, but their origin is one of the neat things about soil chemistry.

Mottles form because reduced iron, which occurs in saturated conditions, is more soluble than oxidized iron. Any areas in the soil with bigger pores, like earthworm burrows, root channels or coarse sand, drain first and become aerobic. These areas accumulate oxidized iron (rust) as it precipitates out along the pores. The diffusion of reduced iron from the surrounding soil provides a steady supply to allow the bright red stains to accumulate.

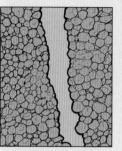

All the pores are filled with water in a saturated soil, and most of the iron is in the reduced form (which is soluble).

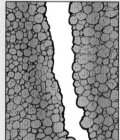

As air enters the largest pores first, the iron converts from the reduced to oxidized form on contact with the air. The oxidized iron precipitates out of solution.

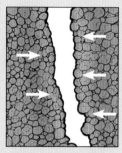

Reduced iron diffuses from the surrounding soil to replace the iron that has dropped out of solution, where it is then oxidized.

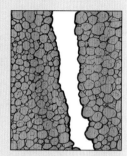

Oxidized iron accumulates along the pore in a reddish mottle.

state (rust), which gives the soil a reddish or yellowish color, to a reduced state, which has much duller colors. When you see these colors in the soil, particularly near the surface, your options are to limit your choice of plants to those that like wet feet or to do something to dry out the soil.

Soil Classification and Mapping

Much of North America—at least the areas with agricultural potential—has been surveyed, and the various soil types have been classified and mapped. These soil maps, available from local departments of agriculture, are a tremendous source of information. They are also intimidating, since they are filled with terminology whose roots are either Russian or Greek.

There is little risk that you will confuse the terminology used in pedology with anything you'll hear in casual conversation. Which is really too bad, since the underlying concepts are often quite straightforward, and it does make sense to differentiate between a prairie soil and a forest or desert soil.

For the casual user of a soil map, however, there is much information to be gained even without learning Russian or Greek. A soil map for your area is covered with squiggly lines that enclose map units, each with a cryptic code. A legend links each of the codes with the name of a *soil series*, along with some other information. The series name, something like "Fox Loamy Sand," includes an identifier for the series ("Fox," in this case) and usually the texture of the surface soil. In the report that accompanies the map is a list of the characteristics of that soil series: the types and depths of horizons in the soil profile, the variations in texture, the drainage class and the presence of other features, like rocks or stones. There may also be information about the limitations of the soil for a particular use and suggestions for suitable crops or ways to improve productivity. The description includes the soil order to which the soil series belongs (see page 53). Details such as the slope of the soil or the presence of shallow bedrock are also provided—information that is obvious if you already own the property but that could be very useful if you are scouting out a new property.

The limitation to soil maps is that they cannot show all the details of the various soil types in a region. As a result, these maps often do not match exactly with what is found at a particular spot. Nevertheless, soil maps are an excellent resource if your goal is to get a sense of the dominant soil characteristics in your area.

How Humans Have Changed Soils Over Time

There is a long history of human intervention in soil-forming processes. Perhaps the earliest example was the periodic burning of grassland areas to disrupt the natural succession of plant species that would otherwise eventually result in unbroken forest. Our distant ancestors conducted controlled burns to create new habitat for the game animals that preferred open grassland and to make hunting easier. Burning also influenced soil formation in the area, creating soils that were somewhere between the true prairie soils and the soils of the surrounding forest.

With the development of agriculture, our intervention in soil-forming processes became more obvious. As we turned over the soil to plant crops, the surface layers of the soil were mixed into a plow layer, or an Ap horizon. That also opened the soil to increased rates of erosion. Many areas were completely denuded of topsoil through poor farming practices. In others, the impact was less extreme, but it did result in dilution of the organic-rich Ah horizon with poorer-quality subsoil.

Agricultural and urban activities have also changed drainage patterns, damming up rivers in one area while draining wetlands in another. In some cases, reshaping the landscape has meant stripping off the topsoil completely so that what is left behind has only a few remnants of the soil's original character.

Advantages and Challenges With Different Soil Types

To many readers, the discussion of soil classification may seem academic, with little relevance to day-to-day life. If you're thinking about the abstract discussions in which pedologists seem to delight, arguing over the distinction between a Brunisol and a Luvisol or a Mollisol and an Alfisol, I would be inclined to agree.

However, it's also true that understanding the fundamental differences between various soil types will help you sort out which plant species will do well or poorly in a particular area. It will also help you determine which soil-management techniques are likely to be needed or effective. This is far more important than knowing the name of a soil order, unless you feel a compelling need to impress a group of pedologists at a cocktail party.

The table on page 53 outlines the dominant pros and cons for each of the soil orders as a medium for growing plants. Not all soils within an area classed as a particular order reflect these exactly, but that is where your diagnostic skills come into play.

For More Information

A number of photo collections available on the Internet illustrate the different soil profiles. These resources include descriptions of many of the physical characteristics of the various soil types, useful information when deciding which management is likely to be needed. Among the best for a quick introduction are the following:

- SoilWeb—University of British Columbia (Canadian soil orders) www.landfood.ubc.ca/soil200/index.htm

- USDA-NRCS Distribution Maps of Dominant Soil Orders http://soils.usda.gov/technical/classification/orders

To dig deeper into soil taxonomy and soil profile descriptions, check out:

- Soils of Canada—horizon descriptions www.soilsofcanada.ca/soil_formation/horizons.php

- USDA Field Book for Describing and Sampling Soils lawr.ucdavis.edu/faculty/gpast/hyd151/soilsfieldguide.pdf

- USDA-NRCS Soil Taxonomy soils.usda.gov/technical/classification/taxonomy

The Root of the Matter

Pedologists (scientists who study the formation and distribution of different soil types) typically hold strong opinions and have a deep sense of history. Canadian pedologists adopted the Russian terms originally proposed by Dokuchaev for most soils (there are some that were not originally identified in Russia), while American pedologists rejected these in favor of building new names from Greek roots.

Dominant Soil Characteristic	Canadian Soil Order	USDA Soil Order	Advantages	Challenges
Newly formed soil	Regosol	Entisol	Minerals present in the parent material haven't been leached out	Low organic matter, may have poor water-holding capacity
Young, partially developed soil	Brunisol	Inceptisols	Minerals present in the parent material haven't been leached out	Soil structure may be poorly developed
Acidic, evergreen forest soil	Podzol	Spodosol	Generally well drained, suitable for acid-loving plants (e.g. blueberries)	Needs amendment with lime for most plants; tendency to low fertility
Deciduous forest soil	Luvisol	Alfisols	Generally neutral to alkaline pH, good moisture-holding capacity	Can have low organic matter; can be susceptible to compaction
Periodically flooded soil	Gleysol	*	High organic matter; very productive if drained	Excess water limits type of crops that can be grown, or reduces yields
Prairie soil	Chernozem	Mollisol	High organic matter, alkaline pH, deep topsoil	Develop in areas that are dry for part of the year, so may need irrigation
Salt-affected soil	Solonetz	*	High organic matter	Extremely sticky when wet, very poor soil structure, excess salts can damage some plants
Desert soil		Aridosol	Generally high in minerals, can be very productive if irrigated	Low organic matter, poor structure
Volcanic soil		Andisol	Minerals present in the parent material haven't been leached out	Thin topsoil, can have excessive drainage
Highly weathered temperate soil		Ultisol	Good moisture-holding capacity	Tend to be acidic; low natural fertility
Highly weathered tropical soil		Oxisol	Good drainage	Acidic, very low fertility, poor capacity to hold nutrients
Soil dominated by expanding clays	Vertisol	Vertisol	High organic matter; alkaline pH; very high capacity to hold nutrients	Sticky when wet; slow to dry out in spring
Muck soils, with more than 30% organic matter	Organic	Histosol	Very high organic matter; easily worked	Nutrient deficiencies (particularly micronutrients)
Soil with permafrost	Cryosol	Gelisol	High organic matter; develop in areas with very long days in summer	Poor drainage; cold soil limits root growth (even with adapted plants)

*not included as soil orders, but as modifiers to other orders (for example, a "Natralboll" would be a salt-affected Mollisol, or an "Aqualf" is a periodically flooded Alfisol)

Soil Texture:
The Bones of Your Soil

Buy land, they're not making it anymore.
— Mark Twain

4

T HE FRAMEWORK UPON which all soil properties rest is the combination of various sizes of mineral particles—sand, silt and clay, along with pebbles, stones and boulders. The proportion of each of the smaller particles (sand, silt and clay) within the soil (for they are always in mixtures, never as just one size of particle) determines the soil texture. This, in turn, has a huge bearing on how the soil behaves—whether it's gritty or sticky, fertile or infertile, excessively dry or overly wet.

Don't confuse soil texture and soil structure, for they are quite different. I have often encountered farmers who speak of managing their soil to change the texture, and I have to suppress the urge to call in the earthmovers and dump trucks. Texture is the mixture of various particle sizes. There are only two ways to change texture; wait through a geological timescale for sand grains to be broken down into finer particles and for chemical changes to convert some minerals into clay; or remove the soil that is there and replace it with something different.

What these farmers mean, of course, is that they plan to change the soil structure, the way the soil particles are combined into larger granules or clods. Soil structure can be changed by the way we manage our soils; soil texture cannot.

We will return to soil structure later, but first, we need to understand a bit more about soil texture.

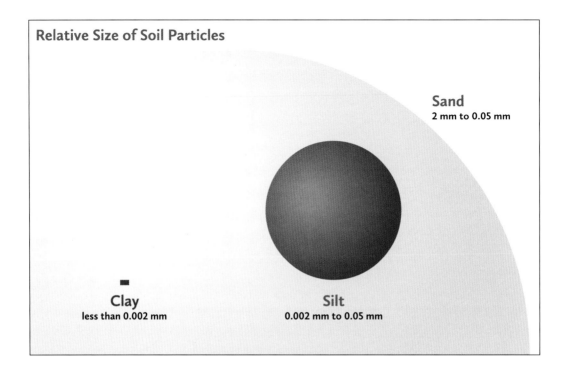

Relative Size of Soil Particles

Sand
2 mm to 0.05 mm

Clay
less than 0.002 mm

Silt
0.002 mm to 0.05 mm

The Components of Soil Texture: Sand, Silt and Clay

The most obvious difference among the various soil particles is their size as shown above, which has a huge impact on how the particles behave, both individually and in combination. Differences in the chemical makeup of the soil particles also play a role in their behavior.

When we divide soils into different texture classes, we consider only particles up to 2 millimeters in diameter, which means the stones in your flower bed are not part of the soil texture. If there is a significant quantity of coarse fragments (pebbles, stones or boulders) in your soil, it may be described as, for example, gravelly loamy sand or stony clay loam. The terms "loamy sand" and "clay loam" describe the texture class. "Gravelly" and "stony" are modifiers which indicate here that the loamy sand has gravel mixed in with it, while the clay loam is mixed with a significant number of stones. Stones or gravel may affect how easy it is to dig the soil, but they have relatively little impact on the way the soil behaves.

Texture classification does not consider any organic matter; it considers only the mineral part of the soil. If you send a soil sample to the lab to have the texture determined, all the organic matter is removed before the measurement is taken.

Sand Grains

Sand grains are visible to the naked eye and the grains feel gritty when you rub them between your fingers. Sand particles are not sticky, so a soil with a lot of sand in it is usually composed of single grains. Because of their large size, sand grains don't pack together easily and the spaces left between particles are relatively large.

Soils with a high sand content drain quickly following rain or irrigation, so sandy soils suffer first during a drought. Very sandy soils also have a poor capacity to hold on to nutrients and are, therefore, often infertile and may be acidic. Because they don't stick together, sand grains are easily detached by strong winds, making sandy soils vulnerable to wind erosion.

Sands can be further subdivided into coarse, fine and very fine size classes. As you might expect, water drains away fastest through the coarser sands, while the finer ones hold more moisture.

Silt Particles

Silt is composed of finer soil particles. Unlike the grittiness of sand, silt feels floury and smooth, though not sticky, and quickly fills up the hollows in your fingerprints. You cannot see silt particles with the naked eye, but they are easily visible under a hand lens or microscope.

Silt is fine enough that its particles pack together very tightly. Indeed, compacted silt can impede drainage. Silt breaks apart easily when tilled, but if there are heavy rains on an unprotected surface, it forms a hard crust through which seedlings have difficulty emerging. Silt is the particle size most easily eroded by water. Its lack of stickiness makes the particles easy to detach, and because of their relatively small size, the particles are quickly carried away by flowing water. (Sand, too, can be detached, but it usually settles out quickly, unless there is a real torrent of water flowing across the soil.)

Clay

Clay comprises the finest soil particles, and the behavior of those particles is very different from that of either sand or silt particles. We refer to clay as a colloid because its particles are fine enough to remain suspended in water for a long time. Clay is what makes soil sticky, partly because of its extremely small particle size and partly because it is made of different minerals than are sand and silt. The family of clay minerals—and there are many different types—is composed of flat sheets, or plates, that stack one on top of the other. These sheets carry a negative electrical charge. Some of that charge is on the surface of the sheets, with more along the edges, so they attract and hold positively charged ions (more on this in Chapter 6). Because this negative charge attracts a thin film of water molecules, clay soil holds on tightly to water.

With its platelike structure and affinity for water, clay absorbs a lot of water between the plates and expands when wet. Then, as the soil dries out, the clay shrinks back to its original size.

You have probably seen photographs of wide cracks in the soil that develop under dry conditions. The degree of swelling and shrinking varies according to the type of clay mineral contained in the soil. Clay minerals change as they weather, losing minerals from within their chemical structure and eventually becoming smaller and thinner. This process takes a very long time. Young clay minerals have the highest mineral content and the greatest capacity to hold on to nutrients, and they undergo the greatest amount of shrinking and swelling. These are found in geologically young areas, where soils have been "recently" deposited by glaciers or by runoff from young mountain ranges such as the Rockies. Soils that have been in place for much longer, like the coastal plain soils in the southeastern United States, are dominated by the more weathered clay minerals (for example, kaolin, which is sometimes used in porcelain because it does not swell when moist). These clay minerals have far less negative charge and therefore don't hold nutrients as well. They also exhibit very little shrink–swell behavior.

Because clay colloids are so fine, there is a lot of space between particles, but these spaces are very, very tiny. The result is that while clay holds on to a lot of moisture, it doesn't let water flow through it very quickly, if at all. Consequently, clay soils tend to remain waterlogged in the spring or after a rain, particularly if they have been poorly managed. Tilling a clay soil when it is wet turns the soil into impenetrable clods that cannot be broken with a hammer when they dry out. On the other hand, the sticky nature of clay makes it difficult for rain or wind to detach particles, so clay is not as easily eroded as is silt or sand.

When we discuss mixed soils later in this chapter, you will see that it takes far less clay content than it does sand or silt to affect soil properties because the fine clay colloids form a film that covers the coarser particles. Beyond a certain point, while there is still a lot of sand and silt in the soil, none of it is exposed. The way that soil behaves is dictated completely by the presence of clay.

Soil Texture: Putting the Pieces Together

Soil with only one particle size does not exist in nature. We are always dealing with mixtures of different particle sizes. Of course, the ideal soil would combine the best aspects of all the particles: the rapid drainage of sand, the friability of silt and the water- and nutrient-holding capacity of clay. That is exactly why we consider a loam soil to be best for gardens, since it does combine all these attributes. In loam, there are equal amounts of sand and silt, but the clay content is far less, ranging between 10 and 30 percent. If there were more than that, the clay would dominate the behavior of the soil.

Beyond the relatively even mixture of particle sizes found in loam soil, some soil textures retain many of the loam characteristics but are combined with the characteristics of the dominant particle size. Thus if there is more sand than silt in the soil, it becomes a sandy loam. If silt dominates, it is a silt loam, and if clay dominates, it is a clay loam. To determine the texture of your soil using a lab analysis that gives the percentages of sand, silt and clay, use the soil texture triangle below.

To explain how to use the soil texture triangle, we'll use the example of a lab analysis that describes the soil texture as

15 percent clay, 30 percent silt and 55 percent sand. The percentages of each particle size contained in the soil are listed along the sides of the triangle. Find the percentage of clay on the left side of the triangle, and draw an imaginary line straight across from there, parallel to the base of the triangle. Then find the percentage of silt on the right side, and draw a line from there parallel to the clay side (i.e., down and to the left). Finally, find the percentage of sand along the bottom of the triangle, and draw a line up and to the left from there, parallel to the silt side. The texture of that particular soil can be identified where the three lines meet. This soil would be classified as a sandy loam.

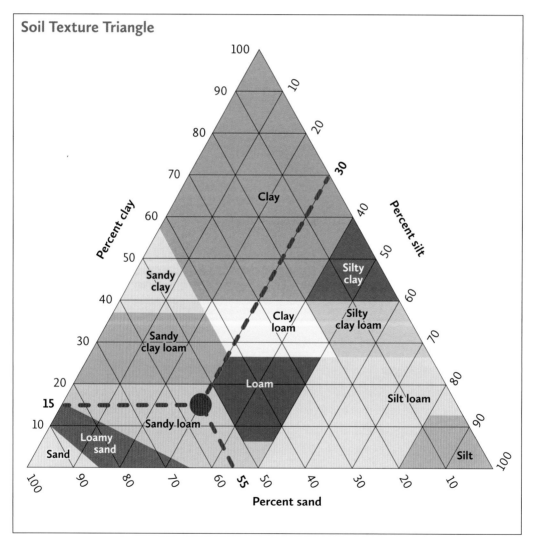

Determining Soil Texture by Hand: A Quick Guide

Experienced soil surveyors become very proficient at estimating the soil texture in the field using only their hands and a bit of water. Most homeowners won't need to develop this level of skill, but it is not difficult to get close enough to the correct texture to help guide your soil management.

Feel Test

Starting with a dry sample, crush a small amount of soil by rubbing it with your forefinger in the palm of your other hand. Then rub some of it between your thumb and fingers to measure the percentage of sand. The grainier it feels, the higher the sand content.

Moist Cast Test

Gradually moisten a small handful of soil, working it thoroughly between your hands to ensure that no dry pockets remain and that the soil is pliable. Do not add so much water to the sample that it gets sloppy. (If this happens, simply add a bit more soil.) Compress moist soil by squeezing it in your hand. If the soil holds together when you open your hand (i.e., forms a cast), pass it from hand to hand—the more durable the cast, the higher the percentage of clay.

Ribbon Test

Roll a handful of moist soil into a cigarette shape, and squeeze it between your thumb and forefinger to form the longest and thinnest ribbon possible. Soil with high silt content will form flakes or peel instead of forming a ribbon. The longer and thinner the ribbon, the higher the percentage of clay.

For more detail on how each texture behaves in these tests, see the table Field Tests for Soil Texture on page 63.

How to Determine Soil Texture by Hand

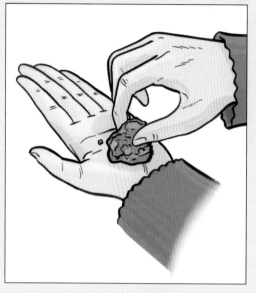

Start with about a tablespoon of soil in the palm of your hand. Gradually moisten it until it is saturated but not sloppy. Work it to break down any aggregates.

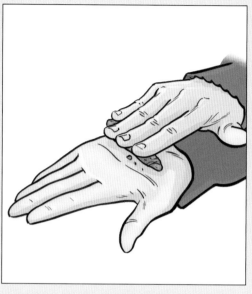

Form the moist soil into a ball to feel how strongly it holds together (the cast test). Roll the ball between your palms to form a thin cylinder.

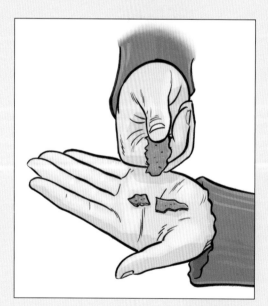

Squeeze the cylinder between your thumb and forefinger to make a ribbon. Short flakes indicate a high silt content.

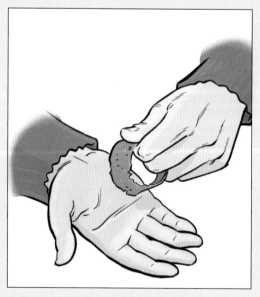

The longer and thinner a ribbon you can make, the more clay your soil contains.

Field Tests for Soil Texture

Texture	Feel Test	Moist Cast Test	Ribbon Test
Sand; loamy sand	Grainy with a little floury material	No cast or very weak cast; does not allow handling	Can't form a ribbon
Sandy loam	Grainy with a moderate amount of floury material	Weak cast; allows careful handling	Barely forms a ribbon: ½–1 inch (1.3–2.5 cm)
Loam	Fairly soft and smooth with obvious graininess	Good cast; easily handled	Thick and very short: 1 inch (<2.5 cm)
Silt; silt loam	Floury with slight graininess	Weak cast; allows careful handling	Makes flakes rather than a ribbon
Clay loam	Moderate graininess	Strong cast clearly evident	Fairly thin; breaks easily; barely supports its own weight
Silty clay loam; silty clay	Smooth and floury	Strong cast	Fairly thin; breaks easily; barely supports its own weight
Clay	Smooth	Very strong cast	Very thin and very long: 3 inches (>7.5 cm)

Managing Each Soil Texture for Best Results

Sand/Loamy Sand

Dominated by coarse particles, these soils dry out and warm up very quickly in the spring. They are easily worked and are generally quite forgiving of cultivation when a little too wet. On the downside, sandy soils don't hold on to a lot of moisture and may need to be irrigated more often than do finer-textured soils. They also have a poor capacity to hold on to nutrients. As a result, it's more effective to make small, frequent additions of nutrients rather than adding a lot at once. Be cautious with fertilizer additions, particularly near the seed, because sandy soils don't provide as much protection against salt injury as do finer-textured soils. If not carefully managed, these soils lose organic matter rapidly.

Sandy and loamy sand soils are ideally suited to plants that do not tolerate wet feet. They are also suitable for early-season crops like lettuce, where rapid warming is a benefit, and in situations where you want to be able to harvest crops, even given a lot of wet weather. They are well suited to perennial crops like asparagus, where rapid drainage helps to protect the plants from root rot and frost heaving.

Sandy Loam

Sandy loam contains slightly more silt and clay than does loamy sand. As a result, it is better able to hold on to water and nutrients but still dries and warms quickly in the spring. Many growers actually prefer a sandy loam to a loam soil, because it can be planted slightly earlier. Usually quite tolerant of cultivation, this soil can become compacted if tilled when moist.

Sandy loam is well suited to most crops, particularly those grown in the "shoulder seasons" (spring and fall). Supplemental irrigation may be needed for plants growing during midsummer. Adding organic matter to this soil, either as an organic amendment like peat moss or compost or by growing green-manure crops, will improve its moisture-holding capacity.

Loam

With a balance of sand, silt and clay, loam soil holds the greatest amount of plant-available moisture. Finer-textured soils hold more total water, but much of it is held too tightly for the plants to access. Loam soil also has good capacity to retain nutrients, but the clay content is not so high that it becomes sticky when wet.

The biggest challenge with loam soil is that it becomes extremely hard and almost impermeable when compacted. Compaction can occur if loam soil is cultivated when wet, particularly if heavy equipment is used. There is not enough clay in this soil to shrink and swell with drying and wetting cycles, so it cannot "heal itself" once it has been compacted.

Loam soil is well suited to most crop types, but some vegetables (e.g., beans and peas) and flowers (e.g., geraniums) grow too well in these soils, generating lots of foliage but not as much fruit or flowers as we might like. This vigorous growth can sometimes lead to fungal diseases infecting the plants under the lush canopy.

Silt/Silt Loam

Dominated by medium-sized particles, these soils are often lumped together with loam soil. They do, in fact, share many of the same advantages, including the capacity to hold both water and nutrients. The biggest difference is that silt provides very little cohesion, which means the soil granules break down easily under rainfall. That can lead to sealing at the soil surface, which can limit water absorption. These soils may also develop a crust that slows down or stops the emergence of young seedlings. They are easily eroded by water as well.

To keep a silt or silt loam soil productive, minimize the amount of cultivation you do (and these soils don't need much) and be sure to make regular additions of organic matter. Planting a fibrous-rooted cover crop (something from the grass family) creates a huge benefit for these soils. Like loam soil, these soils are well suited to a wide range of crops. They are not sticky, which means root vegetables are far easier to clean after harvest.

Clay Loam

In clay loam, the nature of the clay begins to dominate due to the content of fine colloids. When it is wet, it becomes sticky. This soil can be very fertile and has a high capacity to hold on to nutrients. It also provides a large amount of available moisture. When well managed, clay loam has a stable soil structure, but if cultivated or overcultivated when wet, its structure can be broken down to form either hard clods or fine powder that slumps together in the first rain. If a good structure has not been preserved, this soil is slow to drain.

Clay loam is well suited to full-season crops, like potatoes and peppers or late cabbages and broccoli. Early-season crops are generally not a good choice, unless you are imposing specific management techniques to speed up soil warming and drying in the spring (e.g., using raised beds). Rotating crops among different families will help to maintain the productivity of clay loam, particularly if part of the rotation is a green-manure crop.

Silty Clay Loam/Silty Clay

I have heard many complaints about how difficult clay soils are to manage, but I'm convinced that the silty clay loam and silty clay soils are worse. Because of the large amount of silt in these soils, their structure is not as stable as that of clay or clay loam soils. If there is heavy rain before the crop emerges, a hard crust can form on the surface of these soils. They also tend to drain more slowly and therefore take longer to dry out in the spring. They are, however, often high in native fertility, particularly potassium.

The successful management of these soils is largely an exercise in patience. Cultivating and even walking on them when they are wet either glues the soil particles together so that water cannot drain or causes hard clods to form. *Always* wait until these soils are dry before working them. They maintain their productivity best when given a chance to rest, so grow hay crops on them periodically. This approach also works well in large backyard gardens. Plant part of the garden with a mix of clover and grasses, like timothy or tall fescue, leave it for a full growing season, then turn the accumulated forage into the soil.

Clay

The chief advantages of a clay soil are its high capacity to hold on to nutrients and its high buffering capacity to changes in soil acidity. The chief challenges are almost entirely related to the amount of water it can hold. It remains wet and cold in the spring and is slow to drain after a rain. We often refer to clay soil as "heavy," as opposed to the "light" sandy soils. In fact, a dry clay soil weighs less than an equal volume of dry sand, because there are more pore spaces between all the fine clay particles. The difference in weight comes not from the soil itself but from the water that is held in those fine pores. Even a clay soil that feels dry can hold 10 times as much water as can a sandy soil.

Another challenge is the stickiness of the clay when it is wet. Stickiness is an advantage when it comes to holding together soil structure, but if you mess around with clay soil before it is dry, it will stick together in impenetrable clods.

That's the visible damage. More insidious is what happens below the surface. The top of the soil may appear dry, but a few inches down, it is still wet. Cultivation doesn't hurt the top, where it will show, but smears shut the pores at the bottom of the tillage depth. The compacted layer prevents drainage after a rain and blocks the penetration of roots, so the plants are unable to reach the water and nutrients that are deeper in the soil. This can result in weirdly misshapen beets and carrots that have tried to penetrate the hardpan.

As with silty clay loams, patience is key to managing clay soil. Before you head in with the spade or the rototiller, dig down to get a handful of soil from the depth at which you intend to work the soil. Squeeze the soil to see whether it is plastic, like modeling clay, or dry enough to crumble apart.

Clay soil may benefit from tillage in the fall, which allows the soil structure to reestablish itself over the winter as the soil freezes and thaws or wets and dries. A soil that is cloddy in the fall could be quite mellow by spring, as long as you don't mess it up by going onto that soil before it is dry enough. Forming raised beds with clay soil can be very beneficial, both because the beds can drain and warm a bit quicker and because it keeps you from walking over the soil except on the pathways between the beds.

Can You Add Enough Sand to Change the Texture of a Clay Soil?

The short answer is no. Theoretically, it should be possible, but there are a number of practical obstacles. The first is the sheer quantities of material you'd need. Suppose, for example, you had a relatively light clay soil, at 45 percent clay content, and you wanted to end up with a loam texture, at 25 percent clay, by adding washed sand, which presumably has no clay in it. To achieve this change in texture, you would need to add 1,800 pounds (800 kg) of sand to every ton (900 kg) of soil. Put another way, treating a 3-by-10-foot (1 x 3 m) flower bed to a depth of 1 foot (30 cm) for good drainage would require twenty-three 50-pound (23 kg) bags of sand.

The second obstacle is mixing that sand evenly with the clay, which will be extremely difficult, if not impossible. The most likely result is a soil that is still sticky with gritty bits in it.

Finally, adding all that sand will dilute the nutrients and the organic matter in the soil, so the soil flora and fauna will be completely disrupted and the resiliency of the soil reduced. The soil may recover eventually with careful tending, but it will take years rather than weeks.

Organic Soils

Soil texture refers to the mineral part of the soil, but soils that formed in bogs or fens have very high organic-matter content. When the organic-matter content is greater than 30 percent, we assume that the mineral part of the soil is completely covered over with a film of organic material, so the soil no longer behaves like a mineral soil at all. These soils are fluffy and easily worked and provide little resistance to root growth. Therefore, they are often used to grow root crops such as carrots or onions. Because of the way they formed, organic soils start out with very low fertility, but they become very productive with the addition of appropriate nutrients.

While we would consider this type of soil to be ideal for a home garden, it does have its drawbacks. First, organic soils form where it is wet, so if you want to live where these soils have formed, you will be dealing with high water tables and wet basements for at least part of the year. Second, the organic colloids in these soils tend to tie up micronutrients, so to grow plants on organic soils, you must supplement the soil's copper or manganese. Finally, if you buy a bunch of this soil to add to your own, the organic matter will not remain stable in the drier environment of your yard. The added humus gradually breaks down, and the level of soil in your garden will slowly subside as the organic materials disappear.

For More Information

- A good introductory website for soil concepts is members.landscapenl.com/storage/Get%20to%20 Know%20Your%20Soil.pdf.
- You can find a lot of detail about basic soil properties at the **pedosphere.ca** website.

Soil Structure: The Skeleton of Your Soil

To forget how to dig the earth and to tend the soil is to forget ourselves.

— Mohandas K. Gandhi

5

AT THE BEGINNING of the last chapter, I noted the difference between soil texture and soil structure. While texture has a huge impact on the type and stability of the structure that forms, it is ultimately soil structure that dictates how the soil behaves and, in turn, how well your plants grow. Good soil structure creates the ideal balance between large pores that drain freely and let air into the soil and small pores that hold water plants can use.

Soil structure is the way soil particles stick together. Simply put, texture is the bones of the soil, while structure is the skeleton, because it is the way the bones fit together. If there is nothing to keep the particles stuck together, as with a very sandy soil, you have a structureless soil with individual grains. At the other extreme, where the entire soil is stuck together with no fractures or voids, you have a massive soil structure. That's what you get if you cultivate a clay soil when it is too wet, something rice farmers do intentionally in paddy fields to create a soil through which water won't drain. It may be a good strategy for rice cultivation, but it is definitely not what most of us want in our gardens.

If you were to dig down to look at the soil profile in a fencerow, where nothing but grass has grown without any disturbance for many years, or in an old-growth forest, the soil structure would look pretty similar for almost any soil texture. The topsoil would be granular and crumbly, while below that the soil aggregates (discrete clusters of soil particles bound together into larger crumbs, granules or blocks) would be larger, with spaces between them, where water can drain easily. The pattern may change with depth, but because of the combined action of climate, root growth and soil organisms, most native soils have a similar structure near the surface.

When we cultivate the soil, however, its appearance changes, because cultivation breaks down the soil's natural structure. One of the key distinctions among different textures is that some resist the degradation of soil structure better than others. For instance, soils with "sand" in the texture name break down almost immediately, while silt soils hang together a bit longer, though not more than one to two years. The clay soils stay together the longest, but even they can be broken down by continual abuse. You can slow this process—or avoid it completely—by minimizing cultivation and never working the soil when it is too wet.

The stability of the soil structure in your garden is just as important as the type of structure, if not more so. You can "create" structure with tillage by breaking down large clods, but if there isn't enough "glue" to hold the particles together, the soil will break down under the first rainstorm. The cohesion may be provided by clay, but more commonly, it is a mixture of clay and organic matter. In addition, fine roots and fungal hyphae form a lattice structure to hold soil aggregates together. This combination of fibers and glue creates a stable bond, a bit like the glass fibers and resin in fiberglass.

Types of Soil Structure and What They Mean

Granular

The individual soil particles are held together in a loose crumb structure, so there is a mix of large and small pores that let water and air infiltrate easily into the soil and allow roots to grow unimpeded. This is the type of structure you want in your topsoil. A soil with granular structure is sometimes said to be in "good tilth." It generally extends only a few inches down from the surface—deeper in the prairie soils and not as deep in forest soils.

Home Test for the Aggregate Stability of Your Soil

Soil aggregates should hold together when they are wetted. If they don't, the soil will probably crust over when it rains, impeding water or air infiltration and preventing seedling emergence. Lab tests to determine aggregate stability are available, but they are complex and expensive. You can assess this stability at home using a few simple tools.

1. Drop Test

Dig a shovelful of soil, and pick up a clod about the size of a small melon. Drop it from waist height onto a hard surface (a white plastic basin sitting on the ground works well so that you can see the soil particles clearly). Soil with good structure for plant growth will break into small granules but not individual grains. Clods that remain intact or break apart completely indicate poor structure.

2. Watering-Can Test

This test should be performed on a soil that is dry on top and has a granular structure (not crusted). Mark out a 2-by-2-foot (60 x 60 cm) area. Place about 1 gallon (3.8 l) of water into a watering can with a sprinkler head that produces an even pattern of droplets. Sprinkle this volume of water evenly over the marked-out area—it represents half an inch (1.3 cm) of rain, as would typically fall during a heavy thundershower. Compare the appearance of the wetted soil with that of the dry soil adjacent to it. Soils with excellent aggregate stability look the same after wetting as before, while a complete breakdown of the aggregates indicates poor aggregate stability. Most garden soils are somewhere between these extremes.

3. Slake Test

Place about ¼ inch (6 mm) of water in the bottom of a white or clear container. Take an air-dry aggregate from your garden topsoil, about the size of the eraser on the end of a pencil. Place this aggregate in the water, and observe what happens as it absorbs water. The water rapidly entering the pores in the dry soil exerts considerable force on the soil structure. Soils with good aggregate stability will hold together, with only a few grains sloughing off the sides, while soils with poor stability will break down almost completely.

Blocky

As you move deeper into the soil, there is less of the biological activity that creates a granular structure and the aggregates are larger. Typically, a blocky structure has well-defined clods, 2 to 3 inches (5–7.5 cm) across, with obvious fissures between them. These gaps often open during dry weather as the soil shrinks, then close again as the soil is moistened after a rain. Plant roots growing through this zone are concentrated in the spaces between the clods. Although the soil within the clods is largely inaccessible, there is still enough space to allow roots to penetrate deeply into the soil and for excess moisture to drain away.

Columnar

Subsoils with a higher clay content may form into columnar prisms rather than blocks, the result of large shrink–swell cycles as the soil dries and gets wet again. They share many of the characteristics of the blocky soil structure.

Platy

As the name suggests, platy structure has horizontal plates. It is usually a sign of past abuses of the soil and is sometimes found at the surface, where the soil structure has been broken down by heavy rain and has then re-formed in layers that can impede water infiltration and seedling emergence. More commonly, it is found a few inches down in the soil, at the bottom of a layer that has been cultivated in the past. In this case, it is a sign of either compaction or tillage when the soil was too wet. Even a weak platy structure indicates that the current way of doing things is starting to cause problems that will worsen if the abuse continues. Fortunately, the soil generally heals itself with a little help from some deep-rooted cover crops like clover and, of course, by staying out of the garden when it is too wet.

Massive

Massive soil structure occurs when the native soil structure is totally disrupted and re-forms into a solid mass. This soil has suffered severe abuse—the soil equivalent of a person being crushed by an M1 tank. Unfortunately, it is typical of many subsoils in housing subdivisions, where the area has been scraped and reshaped during the development phase and repeatedly driven over by heavy construction equipment in all kinds of weather. Then it is smoothed out to ready it for a thin skin of topsoil.

This massive structure is one of the reasons that so many urban and suburban gardens either fail to thrive or simply die. It is not something that can be fixed with extra fertilizer or irrigation. This soil is almost completely impervious to penetration

Types of Soil Structure

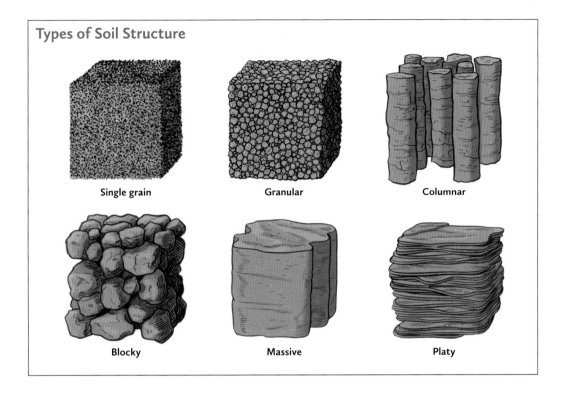

Single grain   Granular   Columnar

Blocky   Massive   Platy

by air, water, nutrients or roots, so plants are unable to reach any added water or nutrients.

Fixing a soil that is this badly damaged will take a lot of time and effort but is a necessary step if your garden is going to be successful. There are two possible approaches:

1. Re-form a granular or blocky soil structure to allow the soil to begin functioning properly again.
2. Build new soil on top of the massive layer so that plants have room to grow without needing to penetrate deeper into the soil.

The first approach is not as simple as breaking the soil into smaller aggregates, although that is part of it. Deep tillage can break up the soil, but it will likely slump together again in a short time if you don't do something to stabilize the structure. These soils are often short of organic matter, so mixing organic materials into the soil will help. Even better is getting some deep-rooted crops growing in the soil. The roots form a scaffolding that holds the new structure together. Adding mulch on the soil surface contributes organic matter and also protects the soil from pounding rain or excess drying. Most important, be sure you do not repeat the mistakes that created this poor structure in the first place.

In the second approach, you're essentially giving up on the soil, at least in the short term. Your goal is to create conditions on top of the abused layer where plants can grow despite the restrictions to deep root penetration. It won't work if you intend to grow trees or other plants that must penetrate deep into the soil to flourish. If you have to dig through this layer later for some reason, you will probably curse it, but it will allow you to get something growing quickly.

The first step is to shape the surface of the massive layer so that water doesn't pool anywhere. The only place for water to go if it accumulates in these areas is up, as in evaporation, a slow process. Next, apply a generous layer of new topsoil—at least six inches (15 cm). This will provide enough depth for the plants to develop a good root system in spite of the restriction on deep penetration. Plants growing in this area need more frequent watering than plants growing in deep soil, because they can't pull water from deeper in the profile. Applying mulch helps to keep the soil from drying out. The roots growing down through the topsoil will gradually begin to penetrate the massive soil and break it down. Eventually, a better structure should reestablish.

Of course, you can use both approaches, which will produce the quickest and longest-lasting response in improved plant growth.

Good Tilth—What Is It and How Do We Manage Our Soils to Get It?

Webster's dictionary defines "tilth" as "cultivated land, tillage" or "the state of a soil especially in relation to the suitability of its particle size and structure for growing crops." It is this second definition that applies when we speak of creating good tilth in the soil. Many adjectives have been used to describe what a soil in good tilth should look and feel like, most of which resonate with gardeners who have worked with soil for many years. But these words may be fairly opaque to those who don't have as much experience. Words such as mellow, friable, crumbly, soft, loose or loamy don't do much to explain what it is.

Part of the problem is that "tilth" takes on different meanings depending on your perspective. Even a soil's "suitability for growing crops" covers a range of different processes and activities. Sometimes we can place the seeds in the ground easily at a depth where they can germinate and grow. For someone using a mechanical planter, tilth can mean that the soil is loose enough for the equipment to penetrate the ground evenly to create a furrow for the seed and then to fall back over the seed to cover it. Or the soil may be of a granule size that allows for good contact between the seed and the soil, and water is able to flow to the seed. It can also mean that plant roots can penetrate easily through the soil to anchor the seedlings and can access water and nutrients for the crop and that the soil is loose enough to allow us to pull weeds and other unwanted plants from the soil without too much difficulty. It's no wonder that it is often easier to say "you'll know it when you see it" than it is to describe good tilth, so perhaps it is best to describe the specific conditions needed for each of the soil functions.

Planting

To plant seed, you could open up a slot in hard soil with a trowel or spade, drop in seed and hope for the best, but this generally does not produce consistently good results. The seed usually ends up being planted too shallow or too deep and doesn't have a covering of soil over top to prevent it from drying out. In friable soil, you can open up a trench in which you can place seeds at the proper depth and cover them over, or sprinkle very small seeds on the surface of the soil and simply firm them into place.

On farms, proper tilth allows the planting equipment to place seed in the ground at the proper depth. Older equipment, in particular, needs a soft, fine seedbed so that shallow planting is avoided and there is soil to cover the seed afterwards. In a garden, this is less likely to be a limitation, as we can compensate for some unevenness in the seedbed with careful cultivation in those spots.

Good Seed-to-Soil Contact

For water to flow to the seed and instigate germination, there must be close contact between the seed and the surrounding soil, as well as pores large enough for water to flow easily through them. Each seed must absorb roughly its own weight in water before it can germinate. If the soil granules are too coarse, there won't be enough points of contact between the soil and the seed. If the soil granules are too fine, there may be lots of contact but the pores are so small that water flows very slowly to the seed. As a rule of thumb, the granules should be one to six millimeters in diameter. Finer seeds should have seedbeds toward the smaller end of this range.

It is important to remember that this is the size range of the soil granules surrounding the seed, but it is not always necessary (or desirable) to have soil this fine at the surface. It may, in fact, lead to problems later if heavy rains cause puddling on the surface and the soil forms a crust when it dries.

In addition to exhibiting the right size of granules, the soil should be firm but not hard, which will allow the pores to form a continuous network to carry water to the seed. Soil that is very loose has lots of pore space, but most of the pores are too large to hold water. This can delay or stop germination completely.

A similar range of soil granule sizes and firmness is desirable for transplants to allow good contact between the soil and the root-ball. Many transplanting failures can be blamed on soils that have been worked too finely, so roots have a hard time penetrating the new soil, or have been left in clods, so water either drains away too quickly or can't move from the surrounding soil into the root-ball.

Root Growth

The health and vigor of the plant you see aboveground are a direct reflection of the health and vigor of the roots growing through the soil. Roots grow, by adding new tissue at the root tips to push through the soil, and this tissue expands in diameter as the roots mature. For that to happen, there must be spaces into which the roots can grow, and the surrounding soil must have enough "give" so that the expanding roots can

Roots that are deformed or suddenly change direction indicate compacted layers in your soil.

push it aside. The roots of plants growing in soil with poor tilth become flattened or deformed as they squeeze past obstructions. These roots are not evenly distributed through the soil but are restricted to the outside of clods or on top of compacted layers. They are not able to reach much of the water and nutrients held in the soil, and thus the plants do not thrive.

At the other extreme, if the soil is too loose, the roots grow easily but are not well anchored in the soil. A strong wind or a dog running through the garden can easily dislodge these plants.

Weed Removal

It is an unfortunate reality of gardening that when growing conditions are right for the seeds you planted, they are also right for weed seeds to germinate. In fact, for gardeners, it is not just death and taxes that are certain but weeds as well. This necessitates pulling or hoeing out weeds so that they don't compete with your garden plants. Weeding is not a particularly difficult task if the soil is in good tilth, but it quickly becomes a chore if the soil is hard and crusted or cloddy. We have all had the experience of seeing weeds we've pulled return a few days later because one little root managed to anchor itself into a clod, allowing the plant to survive and send out new roots.

Building Good Tilth

Creating the physical conditions suitable for plant growth is not as simple as cultivating the ground until the granules have been pounded down to the right size. This approach may occasionally give you the results you want, but that is more good luck than good management, and it probably indicates that the right conditions were already in place before you started working the ground.

The first step to good tilth is to ensure that you have a stable soil structure, which means having adequate organic matter in your soil as well as the right mix of sticky and fibrous material so that the granules hold together. Without this stability, aggressive cultivation is likely to create a mix of clods and powder, as the action of the tillage tool abrades, or wears away, the edges of the clods but doesn't shatter them.

You must always be conscious of the soil moisture. If the soil is too wet, it will smear together as you work it, rather than break into granules. The clods that are formed this way are almost impossible to break apart if there is a high proportion of clay in the soil. Before you start cultivation, always wait until the soil is dry enough that it doesn't form a tight ball when you squeeze it in your hand.

At the other extreme, some soils "set up" if they become too dry and are then difficult to cultivate. These are usually the soils whose organic-matter content is borderline, but they can be cultivated successfully if you are careful.

You can also harness natural processes to help you create good tilth in your soil. Cycles of freezing and thawing and wetting and drying break down clods far more effectively than any mechanical means because this process happens from within the clods themselves. If your soil doesn't easily work down to a nice seedbed, as with many of the clay soils, try working it roughly in the fall and leaving it over the winter. By planting time in the spring, it will be much mellower. Just be sure you don't cultivate it before it is dry enough, or you will undo all the good that has been done over winter.

Tillage: Why and Why Not

The origins of tillage go back to the beginnings of agriculture, as farmers sought better ways to put seeds into the ground. From simple sharpened sticks that created holes into which seeds were dropped, cultivation progressed to wooden plows as early as 3000 B.C. A huge assortment of tillage tools has been developed since then, although often without a clear understanding of the impact they were having on the soil. In the early 18th century, Jethro Tull (a philosopher and the inventor of the seed drill) theorized that plant roots had tiny mouths that took nutrients from the soil as particles. He thought that breaking down the soil into a fine powder would increase nutrient uptake. Even though his theory was incorrect, the system he developed resulted in greatly increased crop yields. Since then, we have developed a far better understanding of the way soil functions and can tailor tillage to the outcomes we hope to achieve, while avoiding some of the pitfalls of too much tillage.

Part of the diversity of tillage tools and tillage systems comes from the many different outcomes that we hope to accomplish with tillage.

Loosening and Aerating the Soil

Almost every tillage implement loosens the soil and opens up larger pores, allowing water to drain out more quickly and air to get in to the soil more easily. Since roots are hidden below ground, it is easy to forget that they need air as well as water. Improving aeration can mean that roots grow deeper into the soil. In turn, the capacity of the root system to absorb water and nutrients is improved. Tillage also creates passages where roots can penetrate the soil more easily, so subsequent tillage operations can be more effective.

The downside to introducing more air into the soil is that it speeds up the decomposition of organic matter. There is a short-term benefit, since decomposing organic matter releases nutrients into the soil that plants can use. But in the long run, the soil has a less stable structure and is more prone to compaction and crusting.

Leveling and Smoothing the Soil Surface

Farmers want a smooth surface on their fields so that their planting equipment can work properly and at a steady pace and seeds can be placed at a consistent depth. Leveled surfaces also enable water to drain off the field freely, which avoids wet patches. While backyard gardeners aren't too concerned about being able to drive across their garden beds, shaping them to control where water does and does not collect is definitely a good idea.

Still, there are limits to the amount of leveling you should attempt by tillage alone. Moving topsoil from knolls and depositing it in hollows can create a smooth and even surface. Unfortunately, it also creates a situation where the topsoil is very shallow or nonexistent on the knolls and very deep in the hollows. Plants trying to grow in the subsoil left where the knolls were may have a tough time from a lack of nutrients or water. Meanwhile, the deep enriched layer in the hollows will subside as the organic matter breaks down, leaving you, once again, with a depression. If you are trying to smooth out undulations of more than a couple of inches, you should seriously consider stripping off the topsoil, leveling the subsoil, then putting the topsoil back in place.

Drying and Warming the Soil

Cultivating the soil to allow more air into it will hasten the evaporation of water. Reducing the water content, in turn, allows the soil to warm up faster. For early planting in the spring, this can be very advantageous. However, there is a risk of damaging the soil structure if the soil is worked when it is too wet. To successfully dry out the soil without causing any harm, "shallow" and "gentle" are the watchwords. Clay soils, in particular, can appear to be dry on top when they are still gummy underneath, so the first cultivation should be only 1 to 2 inches (2.5–5 cm) deep. For small seeded crops, that may be all you need. If you have larger seeds that require deeper planting, you can always do subsequent cultivations after the soil has dried out a bit more.

Improving Water Infiltration

Opening up the soil surface exposes the pores that carry rainwater from the surface into the soil. For that reason, there is better water infiltration on a tilled soil than on one that hasn't been cultivated. But the soil aggregates created by tillage tend to be less stable than the aggregates in an undisturbed soil, so the effect lasts only until the soil is packed down and the pores filled in by the next raindrop impact. After the first good rainstorm, the surface of the soil seals up and more rainwater runs off than if the soil hadn't been tilled.

Incorporating Fertilizer, Manure and Lime

Since plant roots operate below the soil surface, nutrients are more readily available to plants when they are mixed into the soil rather than being left on the surface. Some nutrients can be lost to the air if they remain on the surface. In short, incorporation increases the amount of nutrient that stays around. If you are correcting soil acidity with agricultural limestone, for instance, it must be mixed thoroughly with the soil if it is going to be effective.

Burying Crop Residues

One of the chief goals of the first tillage pass performed on farm fields is to bury the residues left over from the previous crop. These residues can harbor disease spores or insect larvae that could harm the next crop. It is commonly thought that burying the residue reduces the potential damage. While that is true in some cases, the reductions in insect or disease pressure are often less than we might hope.

A more valid reason for making this first tillage pass is that large volumes of crop residue interfere with tillage or planting equipment, particularly older equipment. Newer tillage equipment is designed to allow residue to flow easily under the frame and between the shanks or disks, whereas older machines quickly plug up and drag mounds of residue around the field. Similarly, newer planting equipment is designed to penetrate through crop residue to place the seeds in the soil, where older equipment simply rides up on top of the residue. Neither of

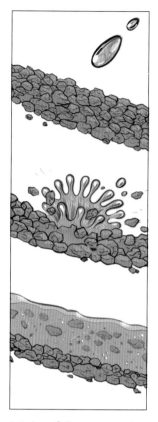

Raindrops falling on exposed soil act like miniature bombs, causing a weak soil structure to be broken down into fine mud that seals up the pores and slows infiltration.

these situations applies in most gardens, but the amount of residue that can be realistically left on the soil surface depends on the gardener's patience—there is no question that last year's plant residue interferes with cultivation by hand.

The unspoken reason for burying all the residue from the previous crop, however, is probably the most powerful. We have been conditioned to believe that a perfect seedbed is completely clean and bare, with no offending pieces of "trash." While I will admit there is a certain aesthetic appeal to neat rows of green shoots emerging from dark brown soil, plants do not necessarily grow any better in these conditions. Indeed, a bare surface may lead to problems with crusting or erosion. It is far better to get used to the idea that crop residue can be a very effective mulch to retain moisture, protect the soil from raindrop impact and increase water infiltration. Try leaving the stalks of your sweet corn after harvest and planting the tomatoes through them. Or plant the corn crop through the vines left over from last year's peas or beans. Just be sure to place the seed into the soil and not directly on top of a piece of residue. It won't work for every crop (small seeded crops like lettuce need unobstructed contact with the soil, for example), but it is effective far more often that its current use would suggest.

Creating a Friable Seedbed

Whether you are planting with a seed drill or by hand, the job is much easier when the soil is soft and loose. Very rarely, you'll encounter soil that is naturally loose and friable (under an old-growth forest canopy, for example). Generally, we need to loosen the soil before we can place seeds in the ground, whether that means working up the entire garden at the beginning of the season or loosening a small area large enough to put a single transplant in the ground.

In either case, you want the soil to be soft enough that you can place the seeds at the proper depth, loose enough that it can cover the seeds, porous enough that roots can grow through it easily and firm but not compacted so that it doesn't dry out completely. The most common mistake is overworking the soil, which destroys its natural structure and results in either an overly fluffy seedbed or one that is too dense. Neither extreme is what you want for a productive garden. The soil should appear granular rather than powdery, with most of the granules at the surface about the size of the seeds you will be planting (the granules are probably smaller below the surface).

Weed Control

Cultivation uproots any weed plants that are already growing and can kill many unwanted plants very effectively. This is especially true with small seedlings that have neither a large root system to hold them in place nor the food reserves to allow them to regrow if they are not completely dislodged. The depth and intensity of cultivation necessary to kill weeds increase with the size of the weeds. Frequent shallow cultivation is most effective for controlling annual weeds.

The downside to tillage for weed control is that it can potentially plant more weed seeds in the germination zone. You should take it as a given that every soil has a bank of weed seeds that has built up over the years and is scattered through various depths. One reason weeds are successful is that deeply buried seeds can remain dormant for years. Mixing the soil with cultivation brings some of those buried weed seeds up to the depth where conditions allow them to germinate, so each tillage pass can plant another crop of weeds. This is why cultivation for weed control should be kept as shallow as possible so that you can exhaust the supply of weed seeds in the germination zone without replenishing the supply from underneath.

Controlling perennial weeds with cultivation is a special case. These plants normally have large and extensive root systems and therefore have a much greater capacity to regrow. It is frustrating to till a patch of weeds, burying the green parts of the plants completely and leaving the soil brown and bare, only to have the same weeds come back as strong as ever a week or two later. Even worse, some perennial weeds can sprout from rhizomes or stems that have been severed from the main part of the plant, so tillage can actually increase the weed population. If you are determined to kill perennial weeds with tillage, the first step is to identify the weed, find out how it spreads and how likely it is to regrow after being uprooted. For species that have deep, spreading root systems and a high capacity to sprout from buried plant parts, it will take repeated cultivation combined with the removal of as much of the root as possible to weaken the plants to the point where they will succumb. In my experience, it is more realistic to expect control rather than elimination of these weeds—you can beat them back to a draw, but you can't expect to win the battle completely.

Tillage Tools

Farmers typically categorize implements as primary or secondary tillage tools. Primary tillage, as the name suggests, is the first pass through the soil that begins the process of loosening hard soils and burying crop residues and other materials. This usually leaves the soil surface quite rough and cloddy and not yet suitable for planting. It may be done in the spring on coarse- or medium-textured soils, but on finer-textured soils, it is generally done in fall to allow the winter to help break down the clods.

Secondary tillage smooths out the rough surface left behind by the first tillage pass. It breaks the large clods into smaller granules that, it is hoped, will provide the proper conditions for a good seedbed. It may take multiple passes, depending on how uneven the soil surface is following primary tillage.

The management of extensive areas of cropland calls for large equipment with equally large tractors to pull it through the ground. In a garden, the necessary equipment is significantly scaled down to suit the size of the area being cultivated, though you are after the same end result. In some cases, garden equipment is just a smaller version of the farm implement. Sometimes the equipment may look quite different. But the key to using any tillage tool effectively, whether power-driven or muscle-driven, is to understand its function. Mechanized equipment gets things done a lot quicker, but manual equipment gives you a lot more control over each tillage operation, allowing you to tailor your approach to specific soil conditions or plant requirements.

A moldboard plow

Primary Tillage Tools

Moldboard Plow

The invention of the self-scouring steel plow by John Deere in the 1830s was a key advance in farming the Great Plains of North America. Described as "The Plow That Broke the Plains," the moldboard plow cut through the tough prairie sod, burying it beneath the surface and leaving the friable topsoil exposed, where it could be worked into a seedbed for wheat or corn. It represented a huge advance over earlier equipment, which either skimmed across the surface without penetrating or became plugged up with sod.

The components of the moldboard plow are a coulter (a straight disk), which cuts through the soil; a plowshare, or share, which creates the furrow by lifting the soil; and a moldboard, which turns each furrow over. If the plow is set properly, each furrow is left leaning up against the next one at a 45-degree angle. Anything on the surface is evenly mixed through the soil by the next tillage pass.

Once the plow has penetrated the soil, the share and moldboard are designed to pull the implement down into the soil so that it doesn't need a lot of weight to penetrate the ground.

If you have bought a country property and want to turn an old meadow into a garden, hiring someone to come in with a tractor and a moldboard plow for the first tillage pass is a very worthwhile investment.

The effectiveness of the plow at soil inversion can lead to challenges. If the implement is set too deep, it will bring to the surface subsoil that is low in nutrients and organic matter. The plow is also a poor choice for tillage in the spring, particularly on clay soils, since it completely disrupts the capillary channels that bring water up from the subsoil to meet the demand of growing crops. Spring-plowed soils can completely dry out to the depth of the plowing, even if there is plentiful moisture in the subsoil, and they don't gain full productivity until a soaking rain wets the entire profile. The down pressure on the bottom of the plowshare can cause significant compaction at that depth, often referred to as a plow pan.

Chisel Plow and Offset Disk

Essentially tractor-drawn heavy cultivators or disks that shatter and mix the soil without inverting it, these implements are very useful where soil erosion is a risk, since the crop residues left on the surface form a protective mulch. Both depend on their weight to penetrate the soil, so there is no garden-sized equivalent. Trying to use a light cultivator or disk meant for secondary tillage will only lead to frustration or broken equipment, as neither is designed to penetrate hard soil.

Shovel or Spade

For the small garden, a shovel can accomplish the same thing as the plow, with the liberal addition of muscle power. It also gives you the chance to literally get a feel for the soil conditions in your garden, and there is certainly a sense of satisfaction as you loosen each shovelful of soil and turn it up against its neighbor.

There is much debate about the proper depth to dig a garden. With soil that is in good condition, there are enough channels and pores to allow deep root penetration, and you probably won't get any benefit from digging any deeper than you need to create a seedbed. Tougher soil conditions might benefit if you

A hand-powered
cultivator

dig farther down in the profile, but I am not a fan of the process
called "double digging" for all cases. This is accomplished by
removing the topsoil and completely turning over the subsoil
to the depth of the shovel while incorporating nutrients and
organic materials. The topsoil is then returned to where it
started, with the addition of fertilizer and compost. While this
practice can certainly create a productive garden, the improvement may not always match the amount of effort required.

Because of the limited amount of soil that each shovelful
can move and the weight of a person compared with that of a
tractor, the risk of damaging your soil by tillage with a shovel is
fairly small. The one thing to avoid, however, is bringing subsoil
to the surface. If you are going to invert the soil as you dig, be
sure to stay within the dark brown topsoil. Do not get down
into the lighter-colored material (the subsoil) below.

A disk

Secondary Tillage Tools

Cultivator

This implement is made up of a series of S-shaped or C-shaped shanks mounted on a rigid frame that is pulled through the soil by a tractor to shatter large clods and level the soil surface. A cultivator is often used to incorporate fertilizer, lime or manure into the soil to about half the operating depth of the implement. In some models, the shanks are mounted in gangs with spaces between so that they can cultivate between rows of crops, like corn or beans, to control weeds.

The biggest risk with a cultivator is operating it too deeply when the soil has not fully dried, smearing the soil below the cultivator tooth and thereby blocking drainage and root penetration. It can also bring ribbons of wet soil to the surface that will dry into hard clods.

Disk

Set to operate at an angle, the curved blades of the tractor-drawn disk aggressively lift, mix and pulverize the soil. The blades may be smooth or notched, depending on the manufacturer and model. The disk moves the soil from side to side, effectively filling in small undulations in the soil surface. It is ideal for incorporating materials that need to be mixed evenly through the soil.

The small disks that are available for garden use do not have enough weight to penetrate hard soil or soil with a lot of crop residue on the surface. Unlike the cultivator or harrow, the disk tends to leave stones buried below the soil surface, where they are less likely to interfere with seeding or harvesting operations.

The biggest drawback to the disk is the weight that is concentrated on the edges of the disk blades as they are pulled through the ground. This can lead to severe compaction if the disk is used when the soil is wet. There is also a danger that the aggressive action of the disk can break down the soil aggregates too finely.

Harrow

The tractor-drawn harrow is less aggressive than other tillage tools and operates no deeper than a few inches into the soil. Mounted on a flexible or rigid frame, the spikes help to break soil aggregates into finer granules and level the soil surface. The harrow is often used as a final tillage pass before planting or to mix broadcast seed into the surface of the soil. It is light enough that it doesn't normally cause much compaction, but there is a risk that overuse will pulverize the soil too finely.

Packer

Also called a cultipacker or roller, this implement consists of smooth or corrugated wheels that are tractor-drawn over the soil to firm the surface and break down clods. Useful in dry soils for helping to bring moisture to the surface through capillary action, it can seal off the surface if it is used when the soil is moist. The result is that rainfall will not soak into the soil very well, and the surface crust that forms may prevent seedlings from emerging.

A chain harrow

Hoe

Depending on how we wield them, the hand tools we use in the garden can do the same jobs as large farm implements. The hoe fills the role of the disk or cultivator, mixing materials into the soil and breaking down large clods. It can also be used to chop residues, to move small quantities of soil (as when forming hills around plants) or to dig furrows in the soil into which seeds can be placed. The flat face of the hoe can also be used as a packer to smooth and firm the soil surface. But as anyone who has tried to use a hoe on hard ground will attest, it is not well suited as a primary tillage tool.

The hoe is available in a variety of sizes and shapes. Some are designed for very specific purposes, such as the narrow-bladed hoes that are built for cutting weeds just below the soil surface. Others have large blades that can be used to chop roots or move substantial amounts of soil. The heavier hoes are useful for specific jobs but may be cumbersome for regular use. The choice of a particular style and size will depend on the hoe's intended

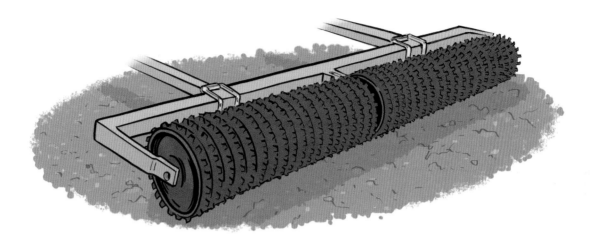

use and what feels comfortable for you. Whatever style you choose, keep the hoe blade sharp with a file or an emery stone. Your back will thank you.

A packer

Rake

The tool for working the soil is the bow rake, with its straight, rigid back, rather than a fan rake, which is used to gather fallen leaves in the autumn. The rigid tines of the bow rake can be used to smooth and level the soil surface and to break up clods, in much the same way that a farmer uses a harrow. The rake also does a good job of distributing material such as compost or peat moss and mixing it into the surface of the soil. For firming soil over a seed row, I prefer the rake to the hoe, since it leaves the soil slightly uneven, which makes it less likely to crust over.

Combination Tools

Rototiller

As the name suggests, the rototiller has tines mounted on a rotating shaft that can chop, mix and pulverize the soil. This machine comes in a range of sizes, from tiny ones for small gardens to large tractor-mounted tillers used by farmers for seedbed preparation. The rototiller is generally the first step into power-aided cultivation for gardeners who have too much area to cultivate by muscle power alone.

A rototiller

With its aggressive action, the rototiller can do both primary and secondary tillage in one pass, breaking up hard soil and chopping and mixing in residues from the previous crop. It is excellent for incorporating manure, compost or fertilizer evenly into the soil to the entire depth of tillage. That said, long stalks or vines, particularly if they are damp or green, can become wrapped around the shaft and reduce the effective length of the tines, which in turn reduces the machine's efficiency. Rocky or stony soils are also a challenge for the rototiller. It doesn't have enough power or strength to smash the stones, but it does operate with enough force for the stones to damage the machine.

In addition, it is easy to work the soil too finely with a rototiller, leaving it subject to erosion or crusting. Even worse is when the soil is too wet at the bottom of the tilled depth. The tines, repeatedly dragged through the soil at high speed, can compact and smear the soil just below the tines, sealing off the soil and preventing drainage of excess water as well as restricting root growth. Be sure to wait until the soil is dry enough to crumble before using a rototiller.

Correcting Common Soil Structural Problems

Compaction

Compacted soils have been squeezed shut so that there is little or no large pore space to allow drainage of excess water, entry of air into the soil or root growth deep into the soil. To correct this, you must reestablish a network of large pores that will remain open. Simply loosening the soil and expecting to have solved all the issues with poor drainage and air entry is unrealistic.

The first step to correcting compaction is to identify how it happened. The second step is to remove that stress from the soil. If the compaction was caused by large equipment during house construction, chances are that it will not happen again. In the case of a compacted area adjacent to a driveway that is in constant use, however, widening the driveway rather than trying to establish a garden may be the best solution.

Once you have eliminated the cause of the compaction, it's time to correct the problem. Loosening the compacted layer is step one. You can use a shovel to break up a compacted layer, but be prepared for some hard work. Work when the soil is moist but not wet. Soil that is wet will smear, and you could actually make the compaction worse.

On the other hand, soil that is both compacted and dry will be extremely hard. You'll need a pickax to break it apart before

This seedling is growing in a very cloddy, compacted soil. All of the roots are confined to a narrow crack, because they cannot penetrate the hard clods.

the shovel will penetrate. To reach the compacted layer, you may have to remove the topsoil and set it aside. You can then dig down through the compacted subsoil, but don't break it up completely or invert the soil. Insert the shovel vertically every 2 to 4 inches (5-10 cm), and wiggle it back and forth to shatter the soil on both sides of the slot. It doesn't hurt if a bit of topsoil trickles down into the slots, as this will form a plane of weakness that roots can exploit later, but don't bother adding a lot of organic material. Replace the topsoil.

Your next goal is to create a stable structure in the soil so that the new pore space will stay in place rather than simply collapsing back to its original state. The best way to do this is to get some roots growing down through these pores to stabilize the new structure. Plant something with deep, vigorous roots, like alfalfa, red clover or oilseed radish.

For areas that are not heavily compacted, you may not need to do any digging if you are willing to leave the area out of production for a year. If plant roots can get through the compacted layer, they will eventually break apart the layer and allow air and water to flow through that zone.

Fragipan Soils

Fragipan soils differ from compacted soils in that they have a subsurface layer which is cemented together by something that has leached down from the surface soil layers. Pedologists argue about the exact processes that form fragipan soils, but the common feature is the presence of a layer that restricts the movement of water and is so hard that roots cannot penetrate it except in isolated cracks. Fragipan soils may be wet in the winter and spring because water is trapped near the surface. Later in the season, they dry out, because water cannot be drawn upward through the cemented horizon to replenish what has been lost at the surface through evaporation.

Even if you were to break up the cemented layer, it would probably re-form within a year or two. The management focus, instead, should be on ensuring that conditions in the zone above the fragipan are favorable for plant growth. Slope the soil surface so that water can drain away rather than accumulate in

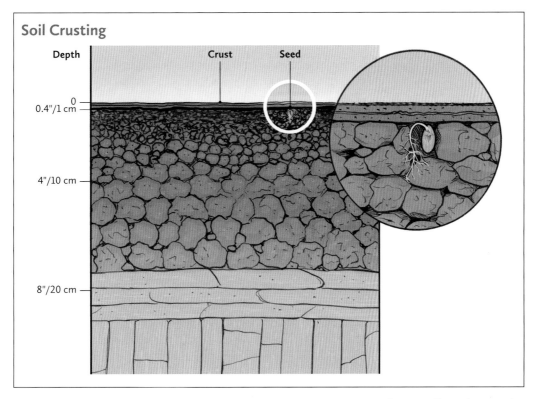

Soil Crusting

Depth   Crust   Seed

0
0.4"/1 cm

4"/10 cm

8"/20 cm

hollows. Maintain good organic-matter levels so that the soil can hold as much moisture as possible. Raised beds are a good option to increase the depth of soil that plants have available for rooting. They also allow the soil in the beds to dry out more quickly in the spring.

Crusts can form when a weak soil structure is pulverized by heavy rain and then dries to the consistency of concrete. Soils high in silt are most affected, but any soil type can form crusts if sufficiently abused.

Crusting

A surface crust forms when the structure of the surface soil has broken down and the soil puddles together under heavy rain. The crust can stop new seedlings from emerging and slow down the infiltration of water. More rainfall runs off the surface crust and less soaks into the soil, where plants can use it. While you can treat a crusted soil to help your plants get out of the ground, you should identify the condition as a symptom of bigger problems and try to prevent it from happening again.

Treating a crusted soil involves gently breaking the crust above the seed row so that seedlings can get through. Using the edge of a hoe, trowel or rake, break up the soil with shallow, vertical strokes. Do not tip the crust up on edge or flip it over, as this could pull out any seedlings that were beginning to emerge before the crust formed and are now trapped within it. This process works quite effectively for large seeded crops that are planted fairly deep, but if you have a shallow-seeded crop in a crusted soil, it might be best to work the area up and start over.

Preventing a crust from forming in the first place is a twofold process:

1. Increase the stability of the soil structure so that crusts don't form as easily;
2. Reduce the energy of the raindrops hitting the soil surface so that they don't do as much damage to the soil structure.

Building soil organic matter by adding manure, compost or peat moss helps improve the stability of the soil structure, as does planting fibrous-rooted plants either as a cover crop or in rotation with the other plants in your garden. Reducing rain impact is not as difficult as it sounds: Simply put something on the soil surface to protect it from the direct impact of raindrops. Farmers do this by leaving crop residue on the soil surface. In gardens, the protective layer is more commonly some type of mulch.

Cloddiness

Soil aggregates that are too large and too hard interfere with good seed–soil contact and with the movement of water and nutrients from the soil to growing plants. You can use tillage to break these clods into smaller granules, but that is usually only partially successful and doesn't address the reason the clods formed in the first place. Like soil crusting, cloddiness indicates a soil with poor structural strength. It may be the result of overworking the soil, which depletes organic-matter levels, or of mixing subsoil in with the topsoil and thereby diluting organic content. As with crusting, the long-term solution is to introduce more organic matter into the soil and to grow fibrous-rooted plants to stabilize the structure.

Short-term fixes for cloddy soil include doing the first tillage of the garden in the fall so that freeze–thaw and shrink–swell cycles can start to break down the clods from the inside. In the spring, wait until the ground is dry enough to crumble before tilling it. Then work the ground only enough to get a seedbed, and resist the urge to take the tiller through for that extra pass.

Fluffy Soils

Soil that is in good tilth should be firm but not hard. Soil that is too fluffy and loose cannot support good plant growth, and while this is usually a temporary condition, it can be just as detrimental to plant growth as soils that are too dense. There is lots of pore space in fluffy soils, but the small pores that can carry water are not connected together, so very little water can reach the germinating seed or the plant roots. The amount of air that can get into these soils, however, means they can dry out very quickly, reducing the available water for plants even more.

A fluffy soil eventually settles back into place, as gravity and rainfall work on it. If seeds have started to germinate in this soil and then dry out, however, this process may not be fast enough to save the crop. Extreme fluffiness is a sign that the soil has been worked too fine. If there is a heavy rainfall, the soil may go from too loose to too tight overnight or the rain can run off, carrying a lot of the loose soil with it and depositing it somewhere else.

The short-term solution is some form of packing implement to firm the soil without compacting it, squeezing some of the large air-filled pores down to the diameter that will hold water. The back of a rake or hoe will do the job for small areas, as will a roller or packer. Concentrate on firming the soil over the seed row and leaving the soil loose between the rows. Weed seeds also have a hard time germinating in fluffy soil, so the crop can get a head start on the weeds. Avoiding a fluffy soil is a matter of limiting the amount of cultivation you do, so beware of overusing the rototiller.

For More Information

The USDA Soil Quality Test Kit Guide provides a comprehensive assessment of the soil, see:
soils.usda.gov/sqi/assessment/files/test_kit_complete.pdf.

Soil Water

Water is life's mater and matrix, mother and medium. There is no life without water.

— Albert Szent-Gyorgyi

6

WATER IS INVOLVED in just about everything that happens in the soil. As it percolates through the earth, it weathers the soil minerals, moves materials down to form soil horizons, dissolves nutrients and carries them to growing plants. The films of water that surround soil particles are home to myriad bacteria and fungi. Provided we manage it properly, the soil provides a reliable reservoir to supply water to our plants.

A 2010 article entitled "Fresh Water" appeared in *National Geographic* with the following introductory statement: "The amount of moisture on Earth has not changed. The water the dinosaurs drank millions of years ago is the same water that falls as rain today." While the claim is a vast oversimplification—water is continually being incorporated into new chemical compounds and then re-formed when those compounds break down—it reminds us that water is an extremely stable compound and that much of the water we use and consume has been cycling through the environment for a very long time.

Plants have an important role in that water cycle. When rain falls to the ground, some of it runs off. Far more, however, soaks into the soil. Once there, it may evaporate directly back into the air or be absorbed by plant roots. Most of the water taken up by plants is just passing through, transpired from the leaf surfaces after it has transported nutrients throughout the plant. This transpired moisture eventually condenses in the atmosphere to form clouds before returning to the ground again as precipitation.

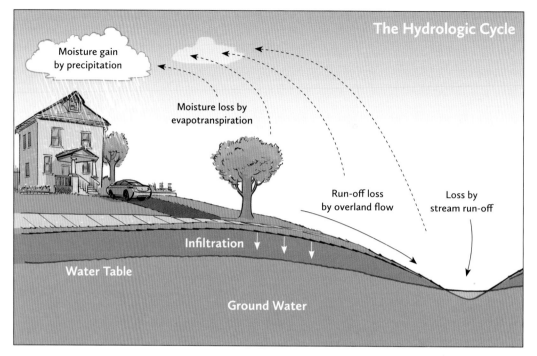

The Hydrologic Cycle

Moisture gain by precipitation

Moisture loss by evapotranspiration

Run-off loss by overland flow

Loss by stream run-off

Infiltration

Water Table

Ground Water

Water continually cycles from the soil and plants, evaporating into the air and then condensing into clouds that deposit rain or snow back on the soil. We affect this cycle by building impermeable surfaces that force the water to run off rather than soak into the soil.

Water taken up by plants arrives through an intricate dance between pore spaces and solid particles. The rain first soaks into the pore spaces in the soil and then flows to where the roots absorb water. Since it doesn't rain every day, water must be stored in the soil, but in such a way that there is space for air in the soil as well. Understanding how these processes work is key to knowing how to get the right amount of water to your plants at the right time.

How Water Behaves in Soil

We've all seen water fall from the sky as rain. It's not hard to picture this water continuing downward once it enters the soil, perhaps at a slower rate. It may be harder to picture that water in soil can also move upward or sideways.

As a result of the negative and positive attraction between water molecules and soil particles, water is held in thin films on the surface of soil particles. This attraction is greatest for the water molecules that are adjacent to the solid surface and

Opposites Attract

Why does water stick to the soil particles? The answer lies in the structure of water (H_2O). The two hydrogen atoms in a water molecule are attached to the central oxygen atom at an angle. If you could see a water molecule, it would look like a shallow "V," with the oxygen at the pointed end and the hydrogen atoms forming the two arms. This arrangement creates a slightly uneven electrical charge, with a small positive charge at the hydrogen end and a small negative charge at the oxygen end. Since opposite charges attract, water molecules tend to stick together.

The same phenomenon causes water to stick to soil particles. The surface of every grain of sand or silt and every clay colloid has a slight negative charge. The positively charged hydrogen end of the water molecule clings to these surfaces, leaving the negatively charged oxygen end sticking out. This, in turn, attracts more water molecules that build up in layers. You can see this effect when water is drawn into a sponge by capillary action. The attraction between the water molecules and the surface of the sponge pulls the water above the surface of the water around it.

is gradually diluted as layers of water accumulate. Since water that is close to soil particles is held more tightly than water that is farther away, the water remains in films over the particles rather than spreading evenly through the soil. When the spaces between particles are small enough, these water films coalesce and small pores become filled with water while the larger pores do not.

Even a very dry soil contains a lot of water (particularly a clay soil), but that water is held in very thin films that are bound tightly to the soil particles. Most of the pore space in dry soil is filled with air. In moist soil, the water films are thicker and are not held as tightly and more of the pores are filled with water, although the largest pores still have air in them. At this point, the attraction of the water to the soil particles is still stronger than the force of gravity.

The way water behaves in the soil in this range of moisture content is counterintuitive. Because the water films are thinner when the soil is dry, the strength of attraction between the soil and the water is stronger. As a result, water in the soil flows from a moist area to a dry area, whether the dry area is below, above or to the side. Water continues to move toward the dry area until the forces holding the water to the soil are equal. This movement may not be very fast, since it depends on flow in thin films over the soil particles and in small pores. Any large void in the soil interrupts this flow or requires the water to make a detour through the surrounding soil. In fact, one of the goals of firming the soil around seeds or young transplants is to ensure that a continuous network of pores is established, allowing water to flow toward the young roots.

When more water is added, more layers of water molecules build up around the soil particles and the attraction between the particles and the outside of the water films becomes weaker. Eventually, the force of gravity is greater than the force of attraction, and water starts to drain down through the soil.

If the soil has a range of textures, the way in which water is held by the soil particles and in pores of different sizes affects the distribution of moisture in the soil. When rain falls on soil whose texture is the same all the way down the profile but does not fall heavily enough to saturate the soil, the water is pulled down into the soil so that all the water films are the same thickness. If the rain falls on soil that has a dry layer of sand above loam beneath the surface, the finer-textured sand pulls more water from the surface soil. By contrast, if a layer of loam soil is over sand, more moisture is held in the loam because there is less attraction for the sandy soil below.

Available Soil Moisture

Water in the soil can be divided into three categories:
1. Gravitational water (water that drains downward).
2. Available water (water that is held in the soil loosely enough that plants can extract it).
3. Unavailable water (water bound so tightly to the soil particles that plants cannot extract it).

The dividing lines between these categories are referred to as *field capacity* and *wilting point*. Field capacity is the point at which water stops draining out of the soil because it is held more tightly than the force of gravity. Think of the soil as a sponge. If you pick up a sponge that has been submerged in a pail of water, water instantly starts to drain from it. Set the sponge on

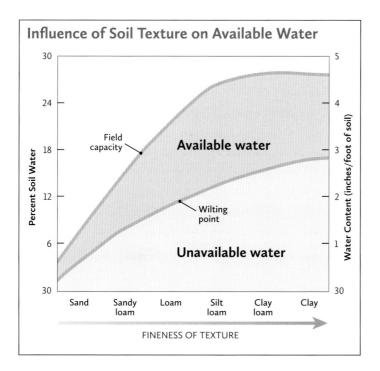

Influence of Soil Texture on Available Water

a coarse screen, and eventually, the water stops draining, but there is still a lot of water within the sponge. At this point, the sponge has reached field capacity. Look closely at the sponge, and you can see that the large holes are filled with air, while the finer pores are still full of water. By squeezing the sponge, you can extract more water from it, but no matter how tightly you squeeze, there is still water left behind.

In the same way, plant roots can pull capillary water from the soil, but some of the water is held in the soil with more suction than the roots can exert. When the soil is dry enough that plants can no longer extract water, the plants wilt, hence the term wilting point. The difference between the moisture content at field capacity and at the wilting point is known as the *available water content* of the soil.

The amount of water that a soil can hold and the amount that plants can use vary with the texture of the soil. Coarse-textured soils, such as sand and gravel, don't hold much moisture at field capacity, because the coarse particles don't have a lot of surface area to which the water films can attach. Plants rapidly deplete this small reserve, so these soils must receive water regularly.

At the other extreme, soils with a high clay content hold a lot of water on the surface of the flat clay minerals and in the tiny spaces between the clay layers. However, much of this water is held so tightly that plants can't pull it out of the soil. A clay soil at the wilting point still contains about two-thirds of the water it had at field capacity. While clay soil has more available water than does sandy soil, it does not have as much as the medium-textured soils. Once silt loams and clay loams have finished draining, they hold almost as much water as does the clay soil, but they don't hold it as tightly, so these soils have the greatest amount of available moisture for plants.

Water Infiltration and Drainage

Before water can be stored in the soil, it must soak into the soil and, at the same time, excess water must be able to drain away. Both of these processes are controlled by the size and arrangement of large pores in the soil. That, in turn, is controlled as much by soil structure as it is by soil texture.

When you water your garden with a hose, three different processes occur. First, the soil absorbs the water very rapidly. Large pores near the surface are filled with air that is easily displaced by the water, then the attraction between the water and the soil particles replenishes the water films around the particles, pulling the water out of the pores.

Soon after, water begins to flow along these films and through fine pores into areas that haven't yet been wetted. This second process occurs very slowly in dry soil but speeds up as the soil gains moisture. Water can move from one soil particle to another only where the water films intersect. As a result, the pathway water takes in dry soil is very convoluted. It must travel around every soil particle to reach the points where the particles touch. As the water films become thicker, the intersection points between particles become much larger and the path the water must follow gets shorter. Some of the small pores become completely filled with water, and the water begins to flow as if it were moving through a pipe. You can see a sharp divide in the soil between moist and dry zones as moisture moves through it; water movement is much faster in wet soil than in dry soil.

Third, as you continue to water, the pore space at the surface eventually fills with water and the soil becomes saturated. At this point, the movement of water through the soil is limited not by the attraction between the water and the soil particles but by how much pressure is pushing the water through the soil and how much space there is for the water to flow through. If water accumulates on the surface, it is being applied faster than the soil can absorb it. If the flow from the hose *is less than* the capacity of the soil to carry away the water, these puddles do not occur.

Among different soil textures and structures, there is a huge range in water flow depending on the total amount of pore space (more pores = more flow), the size of the pores (large pores = more flow) and the connectivity of the pores (more connected pores = more flow). Think of the pore network in the soil as being analogous to the highway system in a big city. The large, straight pores are multilane expressways, the smaller pores are secondary roads, and the fine pores are side streets, alleyways and culs-de-sac. Unlike with city traffic, however, the rule of law with soil moisture is that you can't move onto the faster road until the slower one is completely full. Once the soil is saturated, however, a lot of water is carried through those

Permeability or Porosity?

The words "permeability" and "porosity" are sometimes treated as synonyms, but when used in reference to soil, they mean quite different things. Porosity is the volume of pore space in the soil. Permeability refers to the ease with which water or air moves through the soil and how well the pores are connected. For example, some clay soils have a high degree of porosity because of all the spaces between the fine clay colloids. However, since the pores are so small and do not create a continuous pathway through the soil, these soils are not considered very permeable.

Consider the pads made of closed-cell foam that campers place under their sleeping bags to protect them from the moisture in the ground. While this material is highly porous, it is not at all permeable because the pores are not connected. It has some give, since the air in the cells is compressible, but moisture cannot move up from the damp ground. Anyone who has made the mistake of buying the cheaper open-cell foam pads, which are both porous and permeable, has almost certainly spent a damp, unpleasant night.

Techniques to Improve Water Infiltration

Water that runs off the surface of the soil doesn't do much to meet the needs of the plants growing in that soil. When it rains or we water our gardens, we want as much water as possible to soak into the soil. Here are a few tips to increase infiltration and reduce runoff:

- Keep the soil surface open and permeable. Anything that contributes to creating a stable soil structure helps.

- Avoid working the soil too finely so that it doesn't crust.

- After harvest, leave the remnant stalks and leaves of the crop on the soil surface, where they can protect the soil from the impact of raindrops or the spray from the hose. (Caution: To avoid transmitting disease spores to new seedlings, don't plant the same crop in the same part of the garden the following season.)

- Apply mulch, which performs the same role as crop residues but more uniformly.

- When irrigating, use a gentle spray rather than a stream so that the water does not break down the surface structure.

- Apply water slowly so that it has time to soak into the soil rather than pooling on the surface.

large pores, or expressways; the greater the number of large pores, the faster the water flows through the soil.

Generally speaking, coarse-textured soils have larger pores than fine-textured soils, which allows water to flow through them more quickly. But the dew worm, or night crawler, goes one better, creating vertical channels, or "bio-pores," that carry water from the soil surface much faster than do the pores in the surrounding soil. Dew worms don't particularly like the gritty feel of a sandy soil, so you'll see more worm burrows in silty or clay soils than in sands. If there is a healthy worm population, clay soils can drain water from the surface faster than do sandy soils. Of course, if you break up the wormholes and close the large cracks between clods through aggressive tillage, the rate of drainage declines rapidly.

Water Uptake by Plant Roots

Making sure water reaches the plant is a gardener's main concern. Not only should there be adequate water in the soil, but the roots must be healthy, active and near enough to the water to absorb it.

Active roots require adequate air as well as moisture. Although some wetland plants have developed such unique adaptations as air passages inside the roots, a general rule is that roots do not grow in a saturated soil. If the soil is flooded by a heavy rain, roots quickly go dormant as their oxygen supply diminishes. If the soil is saturated for more than a day or two, roots begin to die, particularly young, tender roots that do most of the water and nutrient uptake. A plant suffering from excess water shows wilting and yellowing—almost exactly the same symptoms as those caused by drought stress.

(There is a large variation among plant species in tolerance to excess water. Check with your local state or provincial extension departments to discuss plant species for your area.)

Since roots avoid growing in saturated soils, plants in areas with a high water table have shallow root systems. Look at the root system of a cedar tree that has blown over in a swamp—the roots extend horizontally for a long way but are only a few inches deep. The same thing happens to annual plants growing in wet soils in the spring. As the ground dries out over the summer, the roots may be unable to grow fast enough to keep up with the dropping water table, and the plants suffer extreme moisture stress.

The other requirement for water uptake is an extensive root system that allows the plant to pull water from a large volume of soil. If the roots cannot grow because the soil has a poor soil structure or is compacted or they are being eaten by grubs or wireworms, the plant's ability to take up water (and nutrients) is restricted.

When it comes to encouraging a large root system, good soil structure is as important as a soil texture with good moisture-holding capacity. It is certainly far more important at the finer end of the soil-texture spectrum. Clay holds on to water very tightly, so the only way to make more water available to plants is to enable roots to reach a larger volume of soil.

Root Systems

Factors That Restrict Root Systems	Potential Solutions
Poor soil structure	▸ Add organic materials to soil. ▸ Plant cover crops or green-manure crops. ▸ Reduce amount of tillage. ▸ Consider changing timing of cultivation.
Compacted soil	▸ Break up compacted layers. ▸ Avoid activities that further compact the soil. ▸ Plant green-manure crops to stabilize soil structure.
Excess water	▸ Direct excess surface water away from garden (e.g., downspouts from eavestroughs, or gutters). ▸ Ensure good surface drainage. ▸ Add subsurface drainage. ▸ Use raised beds to create greater depth of dry soil.
Fragipan (cemented layers)	▸ Use raised beds to increase rooting depth. ▸ Ensure there is good surface drainage.
Insect feeding	▸ Identify pests feeding on roots. ▸ Rotate to different crops in that area. ▸ Use insecticides as a last resort.
Root diseases	▸ Rotate to nonhost crops in that area. ▸ Improve soil drainage or grow plants in raised beds.
Acidic soil	▸ Check soil pH with lab test. ▸ Apply agricultural lime to correct acidity.
Salt injury	▸ Check soil conductivity with lab test. ▸ Reduce or eliminate further applications of fertilizer, manure or compost. ▸ Leach soil with clean water.

Because of the importance of both soil structure and soil aeration, the actual availability of water from the soil may not exactly match the theoretical spread between field capacity and wilting point. Sandy soils, especially coarse sands, have many large pores that carry air into the soil even before the soil has finished draining. Therefore, roots can grow well and extract water from a sandy soil that is wetter than field capacity, creating a situation where coarse sands have more water available to plants than we might predict.

Is Adding Organic Matter Necessary?

Most gardening books emphasize the importance of adding organic matter, which, according to the proponents, soaks up water like a sponge and releases it to plants as they need it.

While organic matter is a central part of soil management, I think its *direct* effect on a soil's moisture-holding capacity has been vastly overstated. It is certain to have an impact in very sandy soils, where the soil simply cannot hold on to water, but loam and clay soils do not share this limitation. Why, then, is organic matter so popular as a means to improve moisture availability in all soil types?

The answer lies in the *indirect* effects of organic matter on soil structure. Organic matter that is integrated into soil covers some of the clay minerals that become sticky when the soil is wet and hard when the soil is dry. It forms a strong glue that prevents the soil granules from breaking down under heavy rain or excess tillage. At the same time, it forms sheaths around the soil granules so that they don't clump together too tightly, allowing the soil to drain more quickly and air to enter into the soil. Organic matter also creates spaces where a vigorous root system can grow.

The situation changes as we move to the finer end of the texture spectrum. Clay soil, particularly if it has a poor structure, has very few pores large enough to carry a lot of air, even when the soil has finished draining by gravity. The absence of air-filled pore space restricts root growth, and plants cannot begin extracting water until the soil has dried out a bit more. As clay soil gets drier, however, it becomes very hard and is difficult for roots to penetrate. This limits the volume of soil from which roots can extract water. So clay soil has less water available to plants than we might expect.

Providing the Right Amount of Water for Plants

Meeting a plant's water needs is always about striking a balance between too much and too little. Adding complexity to the picture, we often grow nonnative plants that are not adapted to our local weather and soil. Therefore, we may have to extensively modify the soil environment to re-create the plant's native conditions so that it can extract the water it needs from the soil.

The right amount of water varies with the season. When we are preparing soil for planting or are harvesting a crop, the soil should be relatively dry so that we don't compact the soil or create hard clods. After planting, however, a good rainfall will help seeds germinate and transplants become established.

Although we can't control when or how much it will rain, we are familiar with local weather patterns. The weather influences the type of soils present in our area and the type of vegetation the annual rainfall supports. If you are in a prairie environment, you expect dry weather in the summer, so your focus is on retaining moisture in the soil. This stored moisture supplies water to your plants during the heat of summer.

In contrast, areas where the native vegetation was once forest generally receive enough rain to supply the needs of the plants during the growing season. Instead of trying to retain soil moisture, we are usually trying to get rid of excess water, particularly at the beginning of the season. (Of course, even these areas face periods of dry weather, when the amount of rainfall doesn't keep up with the plants' needs.)

No matter where you live, the first principle of moisture management is to make sure you have good soil structure. This goes beyond a nice granular structure at the soil surface to include cracks and pores that extend 3 to 4 feet (1–1.2 m) down. Well-structured soil quickly drains away excess water after it rains, allowing air to fill the large pores and provide needed oxygen to the roots. It also has lots of small pores that hold moisture the plant roots can extract, ensuring that roots can exploit the available moisture from a large volume of soil.

Of course, even with good soil structure, we can undermine the ability of plants to utilize soil moisture effectively. The most common mistake gardeners make is watering plants lightly and frequently. This is especially damaging to soils that are already dry. A light watering penetrates only the top half inch (1.3 cm) or so. Roots multiply in this zone to take advantage of the water but are blocked from penetrating any deeper by the dry soil below. At the same time, water cannot move upward by capillary action from deeper in the soil because the dry zone breaks the continuity of the water in the fine pores. As long as you

...our Soil

the plant remains wilted overnight. Leaves begin to yellow or drop off, and the plant is in danger of dying if it doesn't receive water soon.

The irony is that a plant suffering from too much water exhibits very similar symptoms. Leaves wilt, and the plant looks yellow and sickly, but the cause is lack of air to the roots. Without oxygen, the roots shut down and the plant can't absorb water. The wilting leaves really are short of water, but adding more won't alleviate the problem. In fact, watering makes the symptoms worse.

It is important to check the condition of the soil before deciding whether you need to irrigate or not. This is particularly true for transplants, where the roots have not yet grown outside of the root-ball into the surrounding soil. The soil adjacent to the roots may be much wetter or drier than the surrounding soil, so feel the soil right around the roots to be sure there is enough moisture but not too much.

r to suffer, but if the city / for a long weekend, the water needs from stored plants are not equipp... soil moisture.

Your plants are better able to tolerate moisture stress if you water infrequently but deeply. Applying 1 inch (2.5 cm) of water should moisten the soil down to a depth of about 6 inches (15 cm) in a sandy soil or 3 inches (7.5 cm) in a clay soil. This encourages roots to grow to this depth and affords them an opportunity to connect with the moisture deeper in the soil. Once this connection is made, the surface tension of the water pulls the deep moisture upward into the zone where plants can use it.

Too Much Water

Most of us have to deal with a surplus of water from time to time. It may be an annual accumulation of winter snow that melts within a few days, an occasional drenching by a summer thundershower or an ongoing struggle with water that accumulates in a low area and steadfastly refuses to drain away. What you do about it depends on your situation—how often and when rain falls in your area and what you are trying to grow.

Options for Removing Excess Water

Improve Surface Drainage

Part of good garden planning is to create a place for excess water to go so that it doesn't accumulate in a hollow in your yard—unless, of course, you want a bog garden. If your garden is located on a slope, it is easier to prevent water from accumulating, but you may want to shape the ground upslope from the garden to direct excess surface water around the garden instead of through it.

If your garden is in a flat area, especially if it is at the bottom of a slope, you face a bigger challenge. Water that arrives from outside the garden has no place to go. Raised beds are a possible solution, whether permanent or temporary. Permanent raised beds are often contained by a structure that forms part of the garden decor and allows you to achieve a greater depth of dry soil above a wet area, impermeable bedrock or a cemented soil horizon than you could easily accomplish otherwise. And because a raised bed is not walked on, there is less chance of compacting the soil. If the bed is high enough, you can do your gardening without bending down, plus you can concentrate on improving soil within a confined area. Despite these advantages, poor planning can leave you cursing raised beds instead of enjoying them.

Raised beds must be large enough to accommodate the plants you want to grow there. If your goal is to plant a large tree or specimen shrub that won't tolerate wet feet, the bed must be massive, which makes it difficult or impossible to reach the back of the bed without climbing up the wall and tramping on the soil. Remember that retaining walls must be strong enough to withstand the pressure from ice formation as well as the weight of the soil. Poorly constructed walls break down after a year or two, leaving an unsightly and unsafe mess that requires costly repairs. The material used to backfill the bottom of the raised bed must be permeable to allow water to drain away but not so coarse that the bed ends up too dry. Make sure there is a good depth of topsoil over the fill material so that plants have a place to put down roots easily.

Permanent raised beds are well suited to perennials or small trees, where keeping roots up out of saturated conditions in winter is important. You can also use permanent beds for annuals, but it is more common to build new beds for annuals each spring, as the soil is normally tilled at the end of the growing season. Temporary raised beds are created by simply hilling the soil up during spring soil preparation and leaving pathways between each bed to walk on. This not only elevates the bed so that there is a greater depth of dry soil but also increases the depth of the topsoil and concentrates nutrients and organic matter in the zone where the plants are growing.

Make the temporary raised beds narrow enough that you can easily reach the middle from either side but wide enough that the center area can be flat rather than pointed. Beds that are too narrow rapidly dry out, leaving the plants suffering from lack of moisture. Some people circumvent this by burying drip tape within the bed, so the ground dries out quickly in the spring for planting but supplemental moisture can still be added from below the surface during dry spells in the summer. Narrow beds can also break down during the growing season, either from heavy rain or from hoeing and cultivation that disrupt the sides of the hill.

Depending on the size of plants you are growing, various planting patterns are possible: a single row for large, bushy plants, double rows for medium-sized plants or several narrow rows for small plants. You can change the size and arrangement of your beds each year as you rotate crops around the garden or experiment with different layouts. Note that the pathways you have been walking on all summer can become packed and hard, restricting the roots of next year's crop if the temporary bed is relocated on top of this year's path.

Improve Subsurface Drainage

Sometimes, it makes more sense to remove the water below the surface rather than striving to get it to run off. There may not be a place for surface runoff to go, or you may want to keep as much of the rainfall in the soil as you can for your plants to access later.

Water accumulates in an area for two reasons, and solving an excess water problem differs accordingly. First, there may be a high water table fed by water seeping through the soil from surrounding areas. Second, the soil particles may be too tightly packed to allow water to drain.

In the second scenario, the topsoil may be sopping wet while the ground 1 foot (30 cm) below the surface is completely dry. The water can't reach lower due to an impermeable layer. Dealing with this condition is primarily a matter of improving the soil structure to create more large pores to allow water to drain. The poor structure that caused the problem may be a symptom of some other issue, like insufficient humus or cultivating the soil when it was too wet. The solutions for this problem are the same as discussed earlier: the judicious use of tillage, the addition of organic materials and the growth of cover crops and green-manure crops.

If the poor internal drainage is due to a compacted subsoil, you may need to do some deep tillage to break up this layer or plant deep-rooted plants to open channels through the layer to allow water to drain. One of the challenges with this approach, however, is that these plants won't grow deep roots unless the soil is fairly dry. At the same time, the soil can't dry out until the plants have grown the roots that allow the soil to drain. You can see why it is always preferable to avoid compaction, rather than trying to fix it afterward.

Once the soil structure is at the point that water can flow through the soil easily, you can work on improving infiltration of water from the surface. Reducing the amount of runoff keeps

Understanding Tile Drainage

It's a myth that tile drains dry out the soil so that there is less water available to plants in dry weather. Subsurface drains are excellent for removing the excess moisture from spring rains or snowmelt. The soil dries out more quickly and warms up sooner, which allows for earlier planting and quicker germination.

The real mind bender about tile drainage, however, is the fact that removing excess moisture in the subsoil actually increases the amount of moisture available to plants. To understand this, remember that plant roots won't grow into a saturated soil and can't use the water from a soil that is above field capacity, because there isn't enough air for the roots to be active. In an undrained soil, plants use only the moisture in the zone above the water table.

If, for example, a loam soil has 1 inch (2.5 cm) of available water per 1 foot (30 cm) of soil depth and the water table is 1½ feet (45 cm) below the surface, the maximum amount of water plants can use from that soil is 1½ inches (4 cm). Installing tile drains that drop the water table to 3 feet (1 m) below the surface allows the plants to access a total of 3 inches (7.5 cm) of water, because the effective rooting depth has doubled.

rainwater in your own yard; it also lowers the risk of topsoil loss via erosion. Until water has a place to go, however, increasing infiltration simply makes the soil wetter below the surface.

Where the problem is a high water table, you must find a way to drain the water from underneath. Farmers install drainage tiles (perforated plastic or clay tubes) 2 to 3 feet (60–90 cm) below the surface of their fields, with a constant slope toward an outlet into a stream or ditch. It's theoretically possible to use the same strategy in your garden, but only if you can locate an outlet for the drain. Since most cities have bylaws that preclude property owners from hooking into neighborhood storm sewers, this method is easier for rural gardeners.

Change the Plants, Not the Soil

There is a huge range in tolerance to wet conditions among plant species. If your garden is wet, you can save yourself a lot of work and guarantee better success by choosing species that are adapted to wet conditions. This is no different from choosing cultivars that are adapted to survive your area's winter conditions or the amount of sun or shade in your garden. Native species that originated in soil conditions similar to yours are likely to flourish, but if your garden setting is not well drained, avoid plants whose descriptions specify, "Doesn't tolerate wet feet." This affects your choice of trees or perennials more than it does annuals. You may face challenges with plants that need early planting, since the soil takes its time drying in spring and keeps you out of the garden. Cool-season crops like peas or lettuce may not display their optimum quality because they mature when the weather is too hot. Full-season crops may not mature before the first fall frost unless you provide some form of protection.

Too Little Water

In the part of the country where I live, we typically receive lots of moisture over the course of the year, but it doesn't always arrive exactly when plants need it. There is usually a period in midsummer when the rainfall does not meet the water demands for growing crops, and this can last for a month or more. Farther west, the total annual rainfall is much lower, and plants survive based on the amount of moisture stored in the soil. In either case, the capacity of the soil to absorb and hold on to moisture, then release it back to the crop is critical.

You can't alter some of the factors that dictate the ability of the soil to supply moisture, such as soil texture and the weather, but others are quite responsive to management. Even when you can't change the soil, there may be the option of adding water through irrigation, although this is not without its risks. The following outline describes a number of ways to improve the moisture supply in the soil so that irrigation may not be necessary. In the next chapter, we'll look at the pros and cons of various irrigation systems.

Options for Improving Moisture Supply in the Soil

Improve Soil Structure
Building better soil tilth is the first step in improving moisture availability to plants. Improved tilth allows more rainfall to infiltrate the soil. As a result, denser and deeper root systems develop that are able to reach more of the moisture in the soil.

Add Organic Matter

In most soils, the real benefit of adding organic matter lies in the improvement to soil structure. In coarse, sandy soils, however, the ability of organic matter to hold moisture and release it back to plants can make a real difference. For this to be effective, however, you need a lot of organic matter, which may not be practical in a large garden. In this case, your option is to plant green-manure crops, which create organic matter as they grow.

Many of the soils that benefit from added organic matter have very low levels to start with, so they need large quantities. Increasing the soil's organic-matter content from 1 percent to 6 percent takes 22 pounds (10 kg) of dry organic matter per square yard/meter. While this may not seem like much, consider that most composts are 80 percent water, so you need to add 11 pounds (5 kg) of wet material for each 2 pounds (1 kg) or so of dry matter. Once it is in the soil, only about 10 percent of what you add ends up as stable organic matter. Suddenly, we are faced with trying to mix half a ton of organic material into a square yard/meter of soil. Even if this were physically possible, it would create other problems in the soil.

Mulch

Applying a thick layer of mulch over the soil significantly reduces evaporation from the surface, leaving more moisture in the soil for plants to use. It also prevents the surface from crusting over, improving infiltration of water.

Some plants emerge quite well through a thick layer of mulch and tolerate close contact of the mulch with the stalk, but many prefer some bare soil around the base. Before buying a lot of mulch, keep in mind the normal weather patterns for your area and the conditions at the time you apply it. In a wet climate, you probably need more evaporation from the soil surface rather than less, so mulch may not help your garden. If you intend to use mulch as a rescue during a dry spell and the soil has already dried out, the mulch may do more harm than good by preventing moisture from reaching the soil.

The Dark Side of Mulch

You would be hard-pressed to find a suburban perennial bed that doesn't have some kind of mulch on the surface. Many vegetable gardeners have also embraced the mulching movement. But, like everything else in life, it is possible to have too much of a good thing.

Advantages of Mulching	Drawbacks of Mulching
Conserves moisture when conditions are dry	Slows down soil drying in spring or following a heavy rain
Improves infiltration of water into the soil	Absorbs moisture from light showers before it reaches the soil
Suppresses weed growth	Provides habitat for pests that need consistently moist conditions, like slugs
Keeps the soil cool during the heat of summer	Continually moist conditions can encourage fungal diseases of the roots and stalks
Mimics the conditions for many shade-loving species that originally grew on forest floors	

Increase Rooting Depth

Breaking up compacted layers, removing excess water through tile drainage, building raised beds and improving soil structure all help plants develop deeper roots. This automatically increases the depth of soil the roots can exploit and the amount of available water the plants can absorb from the soil.

Plant Selection

Just as some plants tolerate wet soils, other plants thrive under dry conditions. You don't have to seek out plants that are adapted to desert conditions (although you could), but many plants native to the Great Plains have adapted to long periods without rain, going dormant during dry conditions, then springing back to life when moisture is available. These plants do well in most gardens. Plants that depend on deep root systems to tap into subsoil moisture won't grow well where the roots are stopped from growing downward by compacted or cemented layers in the soil.

What About Dust Mulch?

Under dry conditions, cultivating a shallow layer at the soil surface into a fine powder, or dust mulch, slows down evaporation from the surface because the pores in the soil are no longer connected to the pores in the dust mulch and water can't easily be pulled to the surface by capillary action. The dust reflects a lot of sunlight, so the temperature at the base of the dust mulch is cooler than at the surface, further reducing evaporation. During the Great Depression, this saved crops by preserving just enough moisture in the soil, but it also left the soil exposed to the ravages of wind and water erosion. In other words, dust mulching probably has more drawbacks than advantages.

Xeriscaping is a landscaping technique in which plants flourish with limited amounts of supplemental water and involves choosing the proper plants, careful mulching and good soil conditions. It is an excellent gardening strategy where water for irrigation is not available or irrigation is prohibitively expensive. Plants are drought-resistant only when they are well established, however, so you may have to provide extra water to get them started.

For More Information

The USGS Water Science School is a helpful resource: ga.water.usgs.gov/edu.

Irrigation Basics
for Gardens

If gardeners will forget a little the phrase
"watering the plants" and think of watering as a
matter of "watering the earth" under the plants,
keeping up its moisture content and gauging
its need, the garden will get on very well.

— Henry Beston, *Herbs and the Earth*

7

IN AN IDEAL world, soils would hold enough available moisture through the growing season to allow the garden to thrive between the rains that may or may not come every week or so. And, in an ideal world, the squirrels wouldn't dig up my tulip bulbs either. Unfortunately, both scenarios are equally unlikely.

At some point in the growing season, gardeners inevitably have to add supplemental water to make sure their plants flourish. Done correctly, irrigation can mean the difference between plants that thrive during a dry spell and plants that die. Done incorrectly, the process can be expensive, messy and ineffective. Depending on the soil and climate where you live, it can also do long-lasting damage.

To make the best use of your irrigation water, follow these six relatively simple rules:

1. Make sure you have adequate drainage.
2. Apply water according to the plants' needs and how much water is already in the soil.
3. Don't apply water faster than the soil can absorb it.
4. Apply water when water losses through evaporation are minimized.
5. Apply enough water at one time to wet the entire root zone.
6. Plan your irrigation system so that you won't run short of water when you really need it.

The Importance of Adequate Drainage

The key to successful irrigation is good drainage. This may seem counterintuitive: Why would you want to get rid of water when you don't have enough to start with? There are three reasons drainage is a necessary part of an irrigation system. First, soils that are well drained usually have a high permeability, allowing water applied at the surface to reach the roots that need it. Second, since you can't control the weather, it may rain just after you irrigate. If a rainfall saturates the soil because the extra water can't drain away, your plants might suffer. Third, salts build up in the soil when it is watered, and you need good drainage to flush away these salts.

Why Salts Accumulate in Irrigated Soil

Salt accumulation is a common problem in arid and semiarid environments, but it can occur in any environment. While the irrigation water itself may contain salt, the culprits are typically the minerals dissolved by the water as it percolates through the soil. Not all the water is taken up by plants—some of it evaporates directly from the soil surface, leaving behind its load of dissolved minerals, or salts. As more water is pulled to the soil surface by capillary action to replace what has evaporated, the concentration of salts increases. If your household uses hard water, you are probably familiar with the crust that forms on the surface of the soil in which your houseplants grow. Something similar happens around plant roots when they absorb water but leave most of the salts behind. Eventually, the concentration of salts rises to the point where it can injure or kill your plants.

To correct this situation, the excess salts must be leached out of the soil. In moist climates, this generally happens naturally over the winter, but in drier climates, you may occasionally need to apply more water than the plants need to flush out the salts.

Water According to Plant Needs and Soil Supply

To add the appropriate amount of water, it's vital to recognize how much moisture is already in the soil and how much water the plants require.

Generally, large plants need more water than do seedlings, but since they have a larger root system, they are able to pull water from a larger volume of soil. As the temperature rises, evapotranspiration increases, so plants need more water in hot weather. Seedlings use about 0.01 inch (0.25 mm) of water per day, but water requirements increase by up to ¼ inch (6 mm) per day once the crop reaches full growth. One inch (2.5 cm) of available water in the rooting zone represents a 100-day supply at the start of the season but only a four-day supply during maximum growth. Clearly, anything you can do to increase the depth of the rooting zone and the amount of available water will have a huge impact on your plants' ability to survive dry periods.

Plants are most sensitive to moisture shortages during flowering and seed set. Dry conditions before flowering may result in a more compact plant but usually don't significantly affect the number of flowers or the amount of fruit produced. Once the blooms appear, though, the situation changes. Water demand decreases as the seeds are maturing, but for crops such as strawberries or watermelon, in which the goal is to maximize the fruit rather than the seed, water demand stays high until harvest. Too much water at this stage, however, increases the risk of fungal diseases and dilutes the sugars in the fruit, which results in a loss of flavor.

The amount of water available in the soil depends on how much water can be stored in each horizon multiplied by the depth of that horizon added up over the entire rooting depth. With commercial irrigation, the rooting depth and soil texture in each horizon are carefully measured, but few home gardeners need to go this far. It is enough to know that sand doesn't hold much available water, silt loam holds a lot and clay lies somewhere in between.

More important than knowing how much water is available in a soil at field capacity is estimating how much moisture is there now. Observe the feel and appearance of your soil at various depths. Soil with less than 25 percent of the available moisture feels dry and crumbly. Soil that is near field capacity feels moist and leaves an outline of the soil in your hand when you squeeze it. If the soil is above field capacity, you can see water on the surface of the soil when you squeeze it. Be sure to take samples from various depths so that you don't irrigate a soil that appears dry at the surface but holds a lot of available moisture deeper in the soil profile.

Several devices on the market measure the moisture content of the soil. One of these is a tensiometer, a long tube with a porous cup at the bottom and a pressure gauge at the top, available at horticultural suppliers for roughly $100 to $200. As the soil dries, moisture is pulled out through the porous cup, which registers as suction on the gauge. When installed and calibrated correctly, a tensiometer is more accurate than squeezing a handful of soil. To get an accurate measure of the moisture deficit in midsummer, the tensiometer should be installed early in the season. Other moisture meters are based on the electrical conductivity of the soil, which is related to moisture content but is more strongly influenced by the salt content. These typically provide less accurate information than you can gather simply by feeling the soil.

If water is in short supply or you have a large garden, schedule your water applications according to how much water is needed to raise the soil's water content in the rooting zone to field capacity (i.e., the maximum amount of water the soil can hold). There are several ways of calculating this amount, from spreadsheets in which you enter the data by hand to apps that use weather data from the Internet to calculate daily evapotranspiration. Links to a few of these tools appear at the end of this chapter, but you can also check with your local department of agriculture.

Many gardeners base their watering decisions on the look of the soil and the state of their plants, and that approach works in most scenarios. I would make two additional suggestions. First, before you turn on the hose, dig up some soil to see how moist it is below the surface. Second, if the weather has been dry and the long-range forecast is for more hot, dry weather, don't wait until the ground is completely dry before you irrigate. You may find that you simply can't apply enough water to keep up with the needs of the plants, and your garden will suffer.

Slow and Steady

Before plant roots can use it, water has to soak into the soil—runoff water is wasted. Avoid ponding water on the soil surface when you are irrigating. Too much water at once, especially if it is applied forcefully with a hose, breaks down the soil structure at the surface, and the next time you water, the soil is less able to absorb it. Creating a saturated zone at the soil surface also has a negative impact on the biology of the soil and can result in the loss of nitrogen.

The ability of the soil to absorb water is one of the key reasons some soils are better suited to irrigation than others. Sandy soils have low available moisture, so they benefit from supplemental water more frequently than do finer-textured soils; they also have a large capacity for infiltration. Clay soils have more available moisture and don't require irrigation as often, but because the infiltration rate is so slow, it can be a real challenge getting enough water into a clay soil when it does need it.

Minimize Losses Through Evaporation

Sometimes, only a small part of the applied water in irrigation reaches the soil. Much of it can be lost to the air by evaporation. The highest losses are from a fine spray applied during the hottest part of the day. Irrigation is far more effective if you water in the early morning or late evening, when air temperatures are cooler. Use a coarser spray closer to the ground.

An exception to this rule is when we water to cool the plants. Some crops, like strawberries, can suffer from heat stress. Applying a fine mist during the hottest part of the day cools the plants as the water evaporates, in the same way that sweat drying on our skin helps to cool us. But don't apply too much water this way—keeping the plants continuously wet is an invitation to fungal diseases.

Reaching the Root Zone

Frequent shallow watering tends to encourage shallow root systems, which in turn require more irrigation. To avoid this unhappy cycle and to develop healthier plants and root systems, water less often but use more water when you do. This will prompt the roots to utilize more of the natural soil moisture and will reduce the loss of water from the soil surface via evaporation. In other words, more of what you apply reaches the plants. You may be able to water the entire garden at once, then give it a rest for several days. Or you could water different parts of the garden on different days.

Tweaking the System

The weather forecast, whether short or long range, is often wrong. And while you can hope for a certain amount of rain during the growing season, there are no guarantees. This uncertainty is the most common reason to plan an irrigation system so that you'll be ready for the exceptional years.

If you are short on water, switch to a method that uses water more efficiently, such as drip irrigation. Don't let the soil get too dry before you start irrigating. Keep up with plant demand by using a moderate amount of water at one time instead of having to apply a massive amount just to keep the plants from dying. Be conscious of which plants are most sensitive to being short of water, and note their stage of growth. If the sweet corn has just tasseled but the green beans are almost ready to pick, use the water on the corn, because a water shortage at this stage will have an impact on the yield.

Methods of Irrigation

Flood

Flood irrigation is probably the oldest way of getting supplemental water to plants. There isn't a lot of equipment involved, but it can be labor-intensive. Simply pour a bucket of water on the ground around your plants. Make sure the surface of the soil is level and smooth so that the water can spread out evenly.

Because you are putting a lot of water on the surface at one time, there can be excessive losses from evaporation and runoff. With fine-textured soils, this method can seal off the surface, so infiltration of the water is very slow, making the risk of loss that much greater.

Furrow

The next advance that our farming ancestors made was to direct the water into furrows between the plant rows. This method is recommended if you have a ready supply of water that can be run into the garden by gravity. It distributes water to the entire plot and keeps the plant roots up out of the saturated zone. Even so, the water is still on the surface, so this technique has many of the same drawbacks as flood irrigation.

Sprinkler

The most common irrigation method in gardens is some form of overhead sprinkler. This can range from the sprinkler head on a watering can or a hose nozzle to an impulse sprinkler on a telescoping stand that can cover a circle 80 feet (25 m) in diameter. All operate on the same principle—breaking a stream of water into small droplets so that the soil surface is evenly covered without excessive force.

On the plus side, the even coverage of small droplets helps to improve infiltration of the water into the soil. On the negative side, small droplets falling on a large surface area increase the loss of moisture through evaporation. And if the plant tissue stays wet or the spray splashes soil onto the foliage, there is a greater risk of fungal diseases. If possible, water in the morning before it gets too hot so that evaporation is reduced and the foliage has a chance to dry off during the day.

With a handheld sprinkler, you can direct the water to where it will do the most good, focusing on newly transplanted perennials or on the base of individual plants showing drought stress. The two most common mistakes are applying too much water at once so that it ponds on the surface (as with flood irrigation) rather than soaking in and failing to apply enough water. The downside is that it takes time to thoroughly wet the soil. After you've done a quick watering, check to see how deep the moisture has penetrated into the soil. It is probably less than an inch (2.5 cm), when it should be going down 3 to 4 inches (7.5–10 cm).

A stationary sprinkler offers a huge advantage—you can leave it running while you do other things. This sprinkler applies water fairly evenly within its watering pattern, but the pattern often doesn't match the shape of the garden, resulting in some areas receiving either no water or twice the water when the spray overlaps to reach into the corners. A hose nozzle is a good option for the odd corners that are missed by a stationary sprinkler. Whenever you water with an overhead sprinkler, put a straight-sided tin can in the garden so that you can tell how much water you have applied.

An impact sprinkler makes it easy to water a big area at once, but you may have to leave it running over a longer time period because water can flow only so fast through the hose. In general, the larger the area covered, the larger the droplets, so there may be problems with the soil surface sealing from drop impact. The drops are also pushed higher into the air, so a breeze can change where the water lands, which leads to uneven coverage.

Drip

Drip irrigation is the most efficient way to add supplemental water to your soil. It involves a hose or tape that is either perforated with tiny holes or made of a permeable material. The water trickles directly into the ground, right beside the plants that need it. This method practically eliminates losses through evaporation, and there is no water lost between the rows. Another benefit is that plant foliage doesn't get wet, so there are fewer disease problems.

The drip tape or hose can be laid on the surface or beneath the soil, with equal effectiveness. Many gardeners cover the drip hose with mulch so that it can't be seen.

There are a couple of drawbacks to drip irrigation. Because it applies water to a small area, you need a lot of hose or you must frequently move the hose. Commercial operators use drip tape, which is relatively inexpensive per lineal foot even though it needs a special header hose to carry the water to the tapes. It is installed at the start of the season and left in place until the crop is harvested. If you have a big garden, consider this option.

With drip irrigation, you cannot apply a large amount of water at once, so you must start watering while the ground is still moist or you won't be able to catch up with the plant demand for water. This can mean some wasted water if you get an unexpected rain, but that is why you want your irrigated sites to have good drainage.

For More Information

- Irrigation Water Management: Irrigation Methods
  fao.org/docrep/S8684E/s8684e00.htm#Contents
- Irrigation Scheduling Checkbook Method
  extension.umn.edu/distribution/cropsystems/DC1322.html
- Estimating Soil Moisture by Feel and Appearance
  nmp.tamu.edu/content/tools/estimatingsoilmoisture.pdf

A Little Bit of Soil Chemistry

Let me embrace thee, sour adversity,
for wise men say it is the wisest course.
— William Shakespeare, *Henry VI, Part Three*

8

HIGH SCHOOL CHEMISTRY has terrified many students. Aside from the worry that the flame from the Bunsen burner might set something on fire or that the concoction in the beaker might explode, the course work involves abstract theory that doesn't appear to have any connection to real life. In this chapter, we will clarify some of that chemistry and help you appreciate what goes on at ground level in your own garden, allowing you to make better decisions about what additives will help—or harm—your soil.

Whether you use organic or conventional methods, chemical reactions are perpetually taking place in the soil and play an important role in how your soil behaves and how well your plants grow. This chapter focuses on the first of two key processes in the soil that act as controls on all the others: soil acidity and alkalinity. In the next chapter, we will look at the way in which nutrients are held in the soil so that they are available to plants and how this relates to other soil properties.

Why Should I Care About Soil Acidity or Alkalinity?

Pretty much everything that happens in the soil happens within the portion of the soil that is water. This is more properly called the soil solution, since it is not pure water. The composition of this solution depends, in large part, on whether it is acidic or alkaline, which dictates an entire suite of soil properties that determine:

- which minerals dissolve and stay in solution;
- which minerals precipitate out of solution;
- the concentration of nutrients available to plants;
- the stability of some of the minerals that make up the soil;
- the type and number of soil organisms;
- how well particular species of plants will grow in that soil;
- the color of blooms on some flowers; and
- whether a particular type of plant will grow at all.

As this list indicates, it is important to know whether your soil is acidic or alkaline and to understand whether you can change that composition. But first, it is helpful to learn some chemistry shorthand. Instead of saying "acidity or alkalinity," we can encapsulate this concept with the term "pH."

Deconstructing pH

The "p" in pH represents the German word for "power" (*potenz*), and the "H" stands for "hydrogen," so "pH" is an abbreviation for "power of hydrogen," or the concentration of hydrogen ions in solution. In pure water, this concentration is $\frac{1}{10,000,000}$ mols per liter, or 10^{-7}, as expressed in scientific notation. The logarithm of a number expressed this way is simply the value of the superscript (–7); the negative of this number is a positive value (7). As the acidic concentration in the solution increases, this superscript becomes less negative, so the pH gets smaller.

A consequence of this scale is that each unit of pH change represents a 10-fold change in the concentration of hydrogen ions, a difference that continues to multiply as the range gets wider. So instead of two pH units meaning a 20-fold change, it is a 100-fold change; a three-unit change is 1,000-fold, and so on. You can see why it is difficult to make big changes in soil pH.

What Is pH?

Chemists use the term pH as a measure of the acidity of a solution, and it is reported on a scale of 0 to 14. Every solution has both acidic and alkaline, and the balance between them determines the pH. If you add these two components together, you get H_2O, or water. Pure water has a small number of the molecules that have broken into these two ions, but because they are in balance, the pH is neutral (neither acidic nor alkaline). This is indicated by a pH value of 7, right in the middle of the scale. Values below 7 indicate that the acidic ions dominate, while values above 7 indicate more alkaline ions. Think of these two ions on either end of a teeter-totter: As the concentration of one increases, the other decreases, and vice versa. While 7 represents neutrality for many common materials, it's important to note that when discussing soils, a pH of 8.0 is extremely high.

You are probably already familiar with pH without realizing it, since pH is a key component of many food flavors. Acidic foods, such as rhubarb or lemons, are sour to the taste. While there is not as clear a link between alkalinity and flavor, highly alkaline

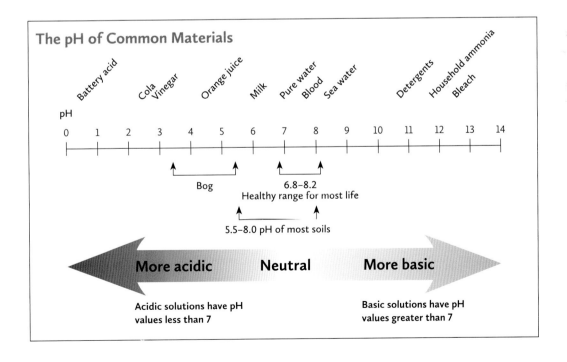

135

Pink and blue hydrangeas

foods may taste salty or bitter, depending on the other components in the food. The pH of a number of familiar materials is shown in the table on page 135.

Soil pH

Soil is more complex than a solution in a beaker, since the ions in the solution are constantly interacting with the solid parts of the soil. Most particles in the soil, both mineral and organic, have a negative charge. Positively charged ions, or cations, like hydrogen move toward and are held by the soil particles, since opposite charges attract (more on this when we talk about cation exchange capacity in Chapter 9). The ions in solution and those held by the soil particles are constantly moving back and forth; as a result, there is an equilibrium between the concentration in solution and on the soil particles. When we add a strong acid to pure water, we can predict how much the pH will change because it will be in proportion to the amount of acid we add. If we do the same thing to a mixture of soil and water, the change in pH is less because the soil particles hold on to the hydrogen ions and take them out of solution. This resistance to change in pH is called buffering. A soil with a high clay or organic-matter content is well buffered, and it takes a large input of acid or alkali to change the pH. Coarse-textured soil with a low organic-matter content is poorly buffered, and its pH changes more easily.

The pH of your soil depends on the type of parent material from which the soil formed, which gives the starting point for pH and buffering capacity, and the amount of acidity that has been added to the soil over centuries.

Type of Parent Material

Since soil minerals come from ground-up rock, the nature of the rock from which your soil formed dictates the starting pH of the parent material. Limestone is alkaline. In the area where I grew up, the parent material was deposited by glaciers traveling over the limestone and dolomite of the Niagara Escarpment, grinding and mixing this material thoroughly with the soil. As a result, the soil's C horizon (deep subsoil) is highly alkaline; even the topsoil has a pH that typically ranges between 7.4 and 7.8.

Rocks dominated by quartz and feldspar, in contrast, are acidic. This includes many of the granites and the associated metamorphic rocks, such as gneiss. As a general rule for igneous or metamorphic rocks (although not necessarily for sedimentary rocks like limestone), lighter-colored rocks form sediments that are acidic, while darker-colored rocks form alkaline sediments. Therefore, soils formed from basalts or micas are alkaline. Rocks formed from sediments follow the rocks from which the sediments were formed, which makes it more difficult to guess the pH tendencies of soil formed from sedimentary rocks.

The type of bedrock that occurs in a geographical region is not always a good indicator of the pH of the parent material, since most soils are formed in sediments that were transported from somewhere else rather than forming in situ. This can mean a significant difference between the pH of the sediment and the bedrock. Near Niagara Falls, for instance, there are areas with only a few feet of soil over limestone bedrock, but the parent material was formed from acidic sediments pushed up by the glaciers from what is now Lake Ontario.

The buffering capacity of the soil depends on the size of the mineral grains and the type of minerals from which it is made. A coarse-textured soil, no matter what its makeup, has a lower buffering capacity than a soil that contains a lot of clay. This lack of buffering comes largely from the relatively low surface area of its large particles, which creates fewer opportunities for reactions between soil and water. In addition, sandy soil is often dominated by quartz crystals, which are almost inert in the soil and don't carry any extra negative charges to hold cations. The net result is that the pH of a sandy soil changes more quickly than the pH of a clay soil. Added acidity in rainfall or from decomposing organic matter pushes the pH down quite quickly, while relatively small additions of lime bring it back up. Remember this when you are deciding how much lime to apply and how often.

One area where mineralogy has an overriding effect on the buffering capacity of soil is in the case of a calcareous soil, which contains calcium carbonate—the same material we use for liming. Acids added to a calcareous soil react with the carbonate minerals that neutralize the lime, and the pH does not start to decline until all the carbonate has been used up.

Acid Additions

Soils naturally become more acidic over time due to the many natural and human-generated processes that add acids to the soil. Natural sources of acidity include root exudates from plants, the breakdown of organic matter, and rainfall, while human-generated sources of acidity include nitrogen fertilizers, acid-forming minerals and acid rain.

Rainfall is naturally slightly acidic, with a pH of about 5.6, because the carbon dioxide in the atmosphere dissolves in water to form carbonic acid. It also contains a small amount of nitric acid (from lightning) and sulfuric acid (from volcanoes).

Plants growing in soil exude acids through their roots. When a root absorbs a cation (pronounced cat-EYE-on) like ammonium, potassium or magnesium, it exchanges the positive charge it absorbs for a positively charged hydrogen ion, which keeps the root electrically neutral. Some plants exude acids from the roots at a high enough concentration to help dissolve minerals in the soil, making nutrients like phosphorus available for the plant to absorb.

Other biological activities in the soil likewise release acids, including the breakdown of organic materials. Organic acids are the result of many digestive processes (whether they occur inside or outside an animal's body). These accumulate in the topsoil and are gradually carried down into the subsoil.

The largest human-generated source of acidity, particularly in agricultural fields, is the addition of nitrogen fertilizers, including manure. Most of this nitrogen is in the form of ammonium, which releases hydrogen ions as it is converted to nitrate. Some fertilizers generate acidity as they interact with the soil, while other materials, such as sulfur, aluminum sulfate and ammonium sulfate, are used specifically to increase the acidity of the soil.

We don't hear as much about acid rain today as we did in the 1970s and 1980s, but a significant amount of acidity in our rainfall is the result of industrial emissions and car exhausts. The biggest component is sulfur dioxide, which forms sulfuric acid in rainwater, along with a smattering of nitric and hydrochloric acids.

Countering the inevitable drive toward greater acidity, a few processes increase soil pH by pulling alkaline materials from the subsoil and depositing them at the surface. Deciduous trees are very effective at drawing calcium and magnesium compounds from the subsoil into their leaves, which are then deposited on the soil surface each autumn. Coniferous trees, which grow naturally in more acidic soils, absorb less of these "basic cations," so the litter on the floor of a coniferous forest tends to be acidic. In places with dry climates, such as the Great Plains, the evaporation that occurs during the year is greater than the rainfall received, so water pulled to the surface by capillary action carries carbonate minerals from deeper in the soil. As the water evaporates, these minerals are left behind in the topsoil.

The net result of these processes is a range of soil pH. You would think we understand these processes well enough to predict the pH range of a particular soil, but that is not the case. On one occasion, I was surprised to find that the soil pH in a soil analysis was acidic, even though most of the soils in the area were alkaline. This soil sample had been collected on a beach ridge left behind by a prehistoric lake, and the soil pH in this narrow band had declined more rapidly than the pH in the surrounding area. The lesson here is to measure the soil pH rather than guessing what it is.

Measuring Soil pH

It is important to know your soil pH because it indicates how the growth of your plants will be affected. When measuring soil pH, there are two main components to consider. The first is the pH of the soil solution itself, which is the environment in which plant roots and soil organisms live. The second is the buffer capacity of the soil, or how difficult it will be to change the soil pH. (We will return to this subject in the section on liming later in this chapter.)

Soil pH can be measured using one of two methods: indicator dyes or electronic meters. Indicator dyes are created to interact directly with the soil solution and are the most accurate method of pH measurement. Remember litmus paper from science class? That's one type of indicator dye. These dyes, which come as a chemical solution that is embedded in paper strips or mixed directly with the soil, change color depending on the pH of the solution with which they come into contact. Soil-testing kits available at garden centers typically include indicator dyes.

Despite its accuracy, however, this method has a few drawbacks. First, the color change usually occurs at discrete pH values, which means you can tell whether the pH is higher or lower than that value but not how much higher or lower. Some provide more precise gradations, but the color gradations can be subtle and hard to discern, particularly if you are one of the 10 percent of adult males who are color-blind. In addition,

Measuring pH in the Lab

A lab pH meter consists of a specialized electrode that generates an electrical potential proportional to the concentration of hydrogen ions around its tip and a voltage meter to measure the strength of this potential. These are calibrated against solutions of known pH, so you can read the pH value directly from the meter. In a lab setting, water or a dilute salt solution is added to the soil to create a slurry. The electrode is then inserted into the soil to make contact with the soil solution. The exact method of making the slurry can affect the pH reading, so the readings for the same soil samples sent to different labs can vary. This difference in readings is usually greater with poorly buffered sandy soils. The measured soil pH from a lab using two parts water to one part soil could be a full pH unit higher than one from a lab using only enough water to saturate the soil. This is not a problem if the lime recommendations are based on the same method the lab uses, but it can cause confusion if you send samples from different parts of your garden to two different labs that don't use identical techniques.

organic matter in the soil can tint the indicator solution and mask the color changes. While relatively easy to use, indicator dyes are not well suited to measuring a large number of samples, which is why they are not used in analytical laboratories.

An electronic meter also has drawbacks. Most meters require that you add water to the soil to form a paste or slurry, which dilutes the soil solution. Since the goal is to measure the concentration of hydrogen ions in the soil solution, this can change the concentration and thus the reading. Some pH meters can be inserted directly into the garden soil, but their accuracy generally leaves something to be desired. The first challenge is achieving good contact between the electrode and the soil solution, especially if the soil is not wet enough. The soil solution must surround the meter sensor, or it may not sense the soil at all.

Analytical labs use pH meters that generate a digital readout of the soil pH. These are usually quite accurate but require careful operation and frequent calibration. With pH meters, you get what you pay for. The pH meters used by analytical labs cost more than $1,000 each. It's unrealistic to expect a device that costs $100 to have the same level of accuracy.

What Should My Soil pH Be?

This is a loaded question. Some gardening books refer to an "ideal" soil pH range, which has created angst among gardeners and resulted in a huge waste of time and money as they chase an ideal value that isn't attainable. There are circumstances in which changing the pH in your soil is desirable, especially with acid soils. More often (particularly with alkaline soils), it is a matter of understanding what you have and accepting that some plants simply aren't going to be part of your garden.

Most plants tolerate a wide range of soil pH, although there are exceptions. The limitations from soil pH normally occur at the low end of the range. It is unclear exactly why acid soils limit plant growth. In some situations, the concentration of hydrogen ions is actually harmful, but this is not a big factor, since plants tolerate much lower pH values in both hydroponic culture and organic soils than they would in mineral soils. Many soil minerals are more soluble in acid soils, so aluminum and manganese could rise to harmful levels. Some nutrient elements are less soluble in acid soils, and nutrient deficiencies may play a role. Rhizobium bacteria, which supply nitrogen to legumes, are very sensitive to acid soils, and plants like beans or peas may do poorly because of a shortage of nitrogen. In most gardens, it is a combination of these processes that causes plants to do poorly.

Plant symptoms occasionally serve as indicators that low soil pH is affecting plant growth. Look for several clues rather than a single "smoking gun." Initially, you may notice that some plants are doing better than others. Plants with the greatest sensitivity to acid soils begin to suffer first, becoming stunted and showing nutrient-deficiency symptoms, such as yellowing leaves or browning of the leaf edges. If you dig up one of these plants, you may see stunted roots with excessive branching. At this point, a soil test is recommended to confirm whether pH is the problem, since other issues can cause similar symptoms.

The popular notion that certain weed species indicate low soil pH has been given far more credence than it deserves. My own experience is that weeds are opportunists, so anything that reduces the growth of your crop allows weeds to grow. Some

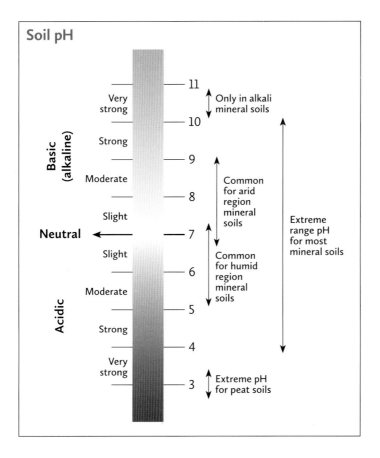

weeds, like sheep sorrel, are far more tolerant of acidic soil than most garden plants and readily occupy the empty spaces left if your peas or corn don't grow due to soil acidity. What the proponents of weeds as indicators of low soil pH fail to realize, though, is that if there are sorrel seeds in the ground, they are just as happy to grow in alkaline soil as in acidic soil once the competition from other plants has been removed.

The literature about weeds as indicators is remarkably inconsistent. One article noted 26 species as potential indicators of acidic soil conditions, then went on to name many of the same species as indicator plants for ideal soil pH and for infertile soil conditions. Weed growth is only one clue as to what is going on in the soil, so add it to your list of possible symptoms, but don't count on it as a reliable indicator on its own.

Optimum pH Range for Garden Plants

Shade and Flowering Trees	
Ash, European mountain	6.0–7.0
Beech, American	5.0–6.5
Birch	5.0–6.0
Hawthorn	6.0–7.0
Holly	4.5–5.5
Honey locust	6.0–8.0
Magnolia, saucer	5.0–6.0
Maidenhair tree	6.0–7.0
Maple	6.0–7.5
Oak, black	6.0–7.0
Oak, pin	4.5–5.5
Oak, red	4.5–5.5
Oak, white	5.0–6.5
Willow, weeping	5.0–6.0
Evergreens	
Arborvitae, American	6.0–8.0
Fir, balsam	5.0–6.0
Fir, Douglas	6.0–7.0
Fir, Fraser	4.5–5.0
Hemlock	5.0–6.0
Juniper	5.0–6.0
Pine	5.0–6.0
Pine, white	4.5–6.0
Spruce, Colorado	6.0–7.0
Spruce, white	5.0–6.0
Yew	6.0–7.0
Vines	
Bittersweet, American	4.5–6.0
Clematis, Jackmanii	5.5–7.0
Honeysuckle, trumpet	6.5–8.0
Ivy, Boston	6.5–8.0
Ivy, English	6.5–8.0
Wisteria, Japanese	6.5–8.0

Vegetables	
Asparagus	6.0–8.0
Beans	6.0–7.0
Beets	6.5–8.0
Broccoli	6.0–7.0
Cabbage	6.0–7.5
Cantaloupe	6.0–7.5
Carrots	5.5–7.0
Corn	5.5–7.5
Cucumbers	5.5–7.0
Eggplant	5.5–6.5
Lettuce	6.0–7.0
Onions	6.0–7.0
Peas	6.0–7.5
Peppers	5.5–7.0
Potatoes	4.8–6.5
Sweet Potatoes	5.2–6.0
Radishes	6.0–7.0
Rhubarb	5.5–7.0
Spinach	6.0–7.5
Squash	6.0–7.0
Tomatoes	5.5–7.5

Ornamental Shrubs	
Azalea, native	4.5–5.5
Cotoneaster	6.5–7.5
Dogwood, red twig	6.0–7.0
Euonymus, winged	5.5–7.0
Fringe tree	5.0–6.0
Heather, Scotch	4.5–6.0
Honeysuckle, Tatarian	6.5–8.0
Hydrangea, PeeGee	6.0–7.0
Lilac	6.0–7.5
Mock orange	6.0–8.0
Mountain laurel	5.5–7.0
Rhododendron	4.5–7.0
Rose, hybrid tea	5.5–7.0
Serviceberry	5.0–6.0
Spirea	6.0–7.0
Sumac	5.0–6.0
Viburnum, double file	6.5–7.5
Viburnum, maple–leaved	4.0–5.0
Wayfaring tree	5.5–7.0
Fruit Plants	
Apple	5.5–6.5
Blueberry, highbush	4.5–5.5
Cherry, sweet	6.5–8.0
Grapes	5.5–7.0
Pear, common	6.5–7.5
Plum, American	6.5–8.5
Raspberry, black	5.5–7.0
Raspberry, red	6.0–7.5
Strawberry	5.5–6.5

What Happens to Plants Outside the Optimum Range?

It is a common misconception—and a source of much unnecessary anxiety—that plants simply won't thrive outside their "optimum" pH range. That can be true if the soil pH goes too far beyond the low end of the range but is generally not the case when it is above the range. Why, then, has this become so entrenched in popular thought?

The short answer is that many of the gardening books have it wrong. High soil pH is not the cause of poor plant growth, but it can be an indicator of other conditions that can limit the growth of some plant species. The availability of some micronutrients to plants, for example, can be caused by high soil pH. It can also be associated with eroded soils, since the subsoil is generally more alkaline than is the topsoil. Plants do poorly in these soils because of low organic-matter content (which limits water availability and soil structural stability) and poor fertility. Lowering the pH of these soils won't overcome the other issues so will have little or no impact on plant growth. Only in rare cases do you need to worry about reducing a high soil pH, and most of the recommended additives aren't effective anyway. It makes sense to avoid adding too much lime to a soil, since that can induce micronutrient deficiencies, but it does not make sense to try to reduce a high pH.

Gardeners living in areas dominated by alkaline soils may look at the optimum ranges of soil pH and decide they need to reduce the pH of their soil. While this response may be justified, it doesn't account for the realities of soil pH. First, the negative impact on plants growing in soil with a pH above the optimum is far less than the impact on plants in soil with a pH below the optimum.

Second, it is extremely difficult to decrease a pH that is too high. It is far more effective to work around the negative effects of the high pH. For instance, phosphorus can get tied up in alkaline soils, so if you are using a high-phosphorus starter fertilizer, apply it in a concentrated band near the seeds or seedlings rather than mixing it throughout the soil. If your soil has a

145

deficiency of a particular micronutrient, spray a dilute solution of that micronutrient on the leaves. Although it won't hurt to mix in moderately acidic organic materials, there will be at least as much benefit from the improved soil structure as from the added acidity.

Acid-Loving Plants

Every rule has exceptions, and while most plants really don't care whether the pH is "too high," there are a few that *must* have acid soil. The best example of this is the blueberry. One of the most frequent questions I have been asked by neighboring gardeners is: "Why aren't my blueberry plants producing fruit?" Since much of the soil where I live is alkaline, these blueberries are being grown in an environment to which they really aren't adapted. Most nurseries do a disservice to their customers by failing to warn them that the promise of abundant crops of sweet blueberries applies only to plants grown in acid soil. Other garden plants that need acid soil to thrive are azaleas and rhododendrons.

Some plants don't need an acid soil to grow but produce different flower colors depending on the soil pH. If you want blue flowers on your hydrangeas, plant them in a soil with a pH below 5.0. At higher pH levels, the blossoms are pink.

It is commonly believed that potatoes grow only in acid soil, but that's not strictly true. Potatoes grow well in alkaline soils, but so does the species of fungus that causes potato scab on the tubers. The fungus is more unsightly than harmful to the plant, but since we prefer our potatoes to look clean, most commercial potatoes are grown in soils with a pH below 5.2.

Correcting Soil Acidity

Acid soil is common across much of eastern North America. Even the drier areas of the Great Plains, which are dominated by alkaline soil, have pockets with low pH. This can limit the

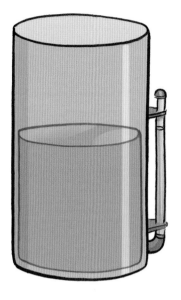

The level of fluid in each tank is the same, in the same way that two different soils have the same soil pH reading. The level of fluid in the smaller tank will show much greater change, however, if the same amount is added or removed from each tank. In the same way, soils have very different capacities to resist change in soil pH, and this is indicated by the buffer pH reading.

growth of many plants but is relatively easy to correct with agricultural lime. The key is knowing where the lime is needed and how much to apply.

To determine whether you need to apply lime, you will have to measure your soil pH. If the pH is below the bottom end of the range for the plants you want to grow, add lime. If it is above the lower end of the pH range for that plant, adding lime is unlikely to make any difference and sometimes reduces the plant's vigor.

A second measurement for the "reserve acidity" in your soil will indicate how much lime to apply. The reserve acidity is the amount of acidity that is held on the soil particles and organic matter and is much larger than the acidity in the soil solution.

The difference between the soil pH and the reserve acidity can be seen in the figure above. Both tanks show the same level of fluid in the sight tube (the small tube at the side of each tank), which is like two soils with the same pH. If you draw the same amount of fluid out of each tank, however, the level drops much more in the small tank than in the large one. In the same way, a soil with low levels of reserve acidity shows greater increases in soil pH when lime is added than does a soil with a lot of reserve acidity. The reserve acidity is related to the buffering capacity of

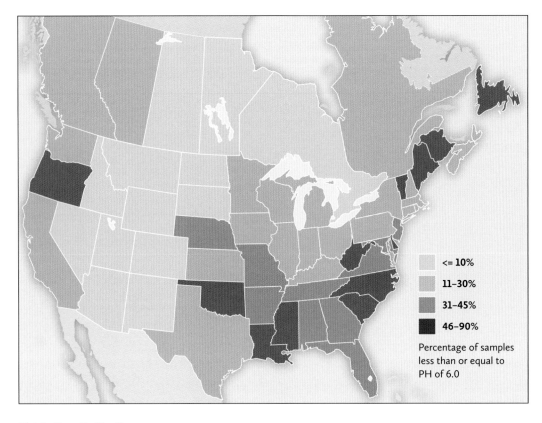

	<= 10%
	11–30%
	31–45%
	46–90%

Percentage of samples less than or equal to PH of 6.0

Distribution of acid soils in North America

the soil. Well-buffered soils have a lot of reserve acidity, while poorly buffered soils have little.

Measuring Buffer pH to Determine Lime Requirement

Several methods are used by soil test labs to calculate the amount of reserve acidity in a soil. The most common is called the buffer pH test. If the soil pH is acidic (usually below pH 6.0), the lab provides a buffer pH along with the soil pH. But a soil test report with two different pH measurements can be confusing, so remember: *Soil pH indicates whether you need to add lime or not; buffer pH indicates how much lime to add.*

The buffer pH reading is used to determine the amount of agricultural lime needed to raise the pH to various levels for each buffer pH value. As the buffer pH gets lower, the lime recommendation increases. The amount of lime is given in tons per acre (or tonnes per hectare), which means little to most gardeners. On a smaller scale, each ton translates into about

¾ ounce per square foot, or 1 pound of lime for 20 square feet (0.45 kg/2 sq m). Because there are a number of different buffer pH methods, be sure the test your lab uses corresponds to the table you are referring to for lime recommendations.

What Happens When You Apply Lime to Soil?

Agricultural lime is composed of finely ground calcium carbonate or sometimes calcium-magnesium carbonate. When applied to the soil, it dissolves and releases calcium and carbonate ions into the soil solution. The carbonate ions combine with the hydrogen ions to form carbonic acid. Because carbonic acid is not stable, it breaks down into carbon dioxide and water. This removes the acidity from the soil solution. The calcium (or magnesium) doesn't play any part in this reaction but is there only as a carrier for the carbonate.

How Quickly Does the pH Increase?

After lime is added, it takes time for the lime to dissolve and interact with the hydrogen ions in the soil solution. As the hydrogen ions are gradually depleted, more are released from the reserve acidity in the soil. If the lime particles are not close to the soil particles, there is an additional delay as each hydrogen ion diffuses through the soil water. You may see some effect on plant growth in the first season following lime application, but it can take up to a year and a half for the reaction to continue to completion. If you think your soil pH is too low, it is best to plan ahead.

You can speed up the reaction (or at least keep it from getting any slower) by using finely ground limestone. Ground limestone should feel floury in your fingers rather than gritty. After applying it on the surface, mix it thoroughly into the soil with a shovel and rake or a rototiller. Agricultural lime is not corrosive, but it can be irritating to exposed skin and mucous membranes. When working with lime, it is best to wear gloves, eye protection and a dust mask.

Is Gypsum an Option?

Because alkaline soil has a high calcium content, it is a common misconception that adding calcium raises soil pH. And since gypsum contains calcium, some gardeners believe it can be used as a liming agent. This is not the case. There is nothing in the gypsum molecule that removes hydrogen from the soil, so adding gypsum does not change soil pH.

What Kind of Lime?

There is much debate about the relative merits of calcitic versus dolomitic lime. The only real difference between them is that dolomitic lime has magnesium while calcitic lime does not. If a soil test indicates that your soil is short of magnesium, use dolomitic lime; otherwise, either one works equally well as a liming agent.

Mitigating Soil Alkalinity

You'll note that we don't say "correcting" soil alkalinity. This is intentional for two reasons. First, reducing the soil pH is not really necessary in most cases. Second, if the soil pH is above about 7.2, it is, for all practical purposes, impossible to lower by conventional means. Most highly alkaline soils contain free limestone as part of their makeup, and all this lime must be eliminated by added acid before the pH starts to drop at all.

If your heart is set on growing blueberries or blue hydrangeas and your soil has an alkaline pH, you may have to take extreme measures. I don't mean adding concentrated acid to your soil or mixing in wheelbarrows of pine needles. I mean digging out the alkaline soil and replacing it with acidic soil. This new soil should contain lots of organic matter to help maintain the low pH by releasing acids as it breaks down. The hole you dig must be large enough to accommodate the entire root system of the plant, with a bit extra around the outside to account for some of the acidity that is neutralized where it contacts the alkaline soil. If there isn't good drainage around the hole, however, even these efforts may not create a suitable environment for your acid-loving plant, because it will wind up sitting in a bowl full of water with no place to go and may develop root rot.

With alkaline soil, the best approach is to treat the problems the alkalinity creates rather than trying to attack the high pH directly. Make sure there is an adequate supply of nutrients, particularly with elements that are less available at high pH, such as phosphorus. To increase availability to the plants, concentrate these nutrients near the plant roots so that there is less contact with the soil. You can accomplish this by placing the fertilizer in a concentrated band near the seed or seedlings, where the plant roots can grow to it, rather than mixing it through the bulk of the soil. The same effect can be achieved for perennial plants by making "pockets" around the plant for the fertilizer or by using the fertilizer spikes that are sold for trees. If the plants are short of a specific element, be prepared to apply micronutrients, such as manganese, as a foliar spray (for help in identifying deficiencies, see diagnostic photos in Chapter 12, Feeding Your Plants.

Reducing Soil pH

In some situations, the soil is acidic but not quite acidic enough to suit the plant species you want to grow. In these cases, it may be worthwhile to amend the soil to reduce the soil pH. Remember, however, that buffering of soil pH works in both directions, so while you may be able to easily drop the pH of a sandy soil, the same rate of amendment on a clay soil will not make any measurable difference.

Numerous materials are promoted to reduce soil pH. While each of them adds acidity to the soil, they all have widely different properties that confer advantages and drawbacks.

The rate of sulfur required to drop the pH from 6 to 5, for example, is in the range of $\frac{1}{10}$ ounce per square foot (1 pound per 160 square feet/0.45 kg per 15 sq m) on a sandy soil and up to $\frac{1}{3}$ ounce per square foot (1 pound per 50 square feet/0.45 kg per 4.6 sq m) on a loam soil. Adding too much sulfur can mess up the chemistry of the soil, allowing sulfides to build up that are toxic to plants and inducing deficiencies of other nutrients. It is better to proceed slowly, adding a bit more the following season, than to apply too much all at once.

Materials for Reducing Soil pH

Material	Advantages and Cautions	Amount to Equal Acid Produced by One Unit of Sulfur
Elemental sulfur (S)	▶ Relatively safe for plants except at extreme rates ▶ Takes 1 to 2 years for full effect	1
Sulfuric acid	▶ Immediate reduction in pH ▶ Requires extreme caution to handle safely ▶ Kills any plant tissue it contacts ▶ Destroys soil organic matter	3
Iron (Fe) or aluminum (Al) sulfate	▶ Rapid pH reduction ▶ Fairly safe for plants except at very high rates ▶ Some risk of aluminum toxicity from aluminum sulfate if used at high rates	9 (Fe sulfate) 4 (Al sulfate)
Ammonium sulfate	▶ Moderately fast pH reduction ▶ Also provides nitrogen to plants ▶ Can cause salt injury to plants if used at high rates	2
Acidic organic materials (peat moss, pine needles, etc.)	▶ Immediate effect, plus a long-term effect from acids released during decomposition ▶ Large volumes of material are required ▶ Effectiveness is extremely erratic, as most of these materials are quite variable	? (my guess would be in the hundreds to thousands)

For More Information

Most of the readily available information on soil pH and liming has been created for use by farmers on agricultural fields. Check with your local department of agriculture for information that is pertinent to your area. Below are some useful resources to get you started.

- *Soil Fertility Handbook*. Ontario Ministry of Agriculture, Food and Rural Affairs. Toronto: Ontario Legislative Library, 2006.
- *Soil Fertility and Fertilizers: An Introduction to Nutrient Management* by John Havlin, John and Samuel L. Tisdale, Werner L. Nelson, James D. Beaton. Toronto: Pearson Prentice Hall, 2005.
- Royal Horticultural Society, United Kingdom
  apps.rhs.org.uk/advicesearch/profile.aspx?pid=239
- University of Massachusetts, United States
  extension.umass.edu/turf/fact-sheets/soil-ph-and-liming
- Rutgers University, New Jersey, United States
  njaes.rutgers.edu/soiltestinglab/pdfs/ph-Lime-req.pdf
- North Carolina State University, United States
  soil.ncsu.edu/publications/Soilfacts/AGW-439-50/
  SoilAcidity_12-3.pdf
- Australia
  dpi.nsw.gov.au/__data/assets/pdf_file/0007/167209/
  soil-acidity-liming.pdf

How Soil Holds Nutrients

Still other seed fell on good soil, where
it produced a crop—a hundred, sixty
or thirty times what was sown.

— Matthew 13-8, *The Holy Bible*

9

A HEALTHY SUPPLY OF nutrients from the soil is dependent on a bit of a balancing act. The soil must be able to hold nutrients tightly enough that they don't all leach away during a heavy rain but loosely enough that the plants can access them. If this weren't possible, the activity of growing plants in soil would look a lot like hydroponic production in a greenhouse, where each of the plants' required nutrients would have to be continually added to the water in proper proportion. Not only would that be expensive, but any small mistake in mixing the nutrients could hurt or kill the plants.

Fortunately, soil does have a way of holding on to and releasing nutrients as plants need them. Every soil particle carries a slight electrical charge on its surface that attracts and holds nutrients with the opposite electrical charge. It's not a strong chemical bond like the one that exists between hydrogen and oxygen in a water molecule, but it allows the soil to behave a bit like Velcro—it holds the nutrients, yet they can still be pulled away. In soil, however, there is always something waiting to occupy any empty space. As a result, there is a continual exchange between the soil solution and the soil particles. For an analogy, imagine the seating at a bar in a busy nightclub—every stool is occupied, and a crowd is waiting. As soon as someone vacates a bar stool, another person immediately claims that place.

The dominant charge on soil particles is negative, so the nutrients that are held by the particles are positively charged ions. Any chemical that carries a positive charge when it is dissolved is called a *cation*. When referring to the ionic form of a chemical instead of the element itself, chemists use a small raised plus sign (a superscript) to indicate that it carries a positive charge. Nutrients that fall into this group include ammonium, calcium, magnesium and potassium. Other common elements in the soil also fall into this group, like hydrogen and sodium.

The quantity of cations that can be held in the soil is set by the amount of negative charge, hence the term *cation exchange capacity*, or CEC. This value is often used as the measure of the fertility of the soil, since soil with a low CEC cannot hold many nutrients. Although soil with a high CEC may hold a lot of nonnutrient cations, it typically holds enough nutrients to ensure that it is more fertile (i.e., better able to supply the nutrients plants need) than a low–CEC soil.

The CEC is dependent on the amount of surface area in the soil and the type of minerals present. With some minerals, as with the quartz in sand grains, the negative charge is very tiny; in addition, coarse sand grains do not have a lot of surface area to carry the charge. Therefore, sandy soils have a low CEC. By contrast, clay soils have a lot more surface area because of the small particle size, and clay minerals carry more negative charge because of their chemical makeup. That said, there is significant variation among clay minerals. For instance, the CEC is lower in the heavily weathered kaolin clays of the coastal plain, in the southeastern United States, than it is in the illite and montmorillonite clays of the Great Plains.

The other source of CEC in the soil is organic matter, particularly the fully decomposed organic part we call humus. This material carries a large negative charge, which means that each gram of humus adds about 10 times more CEC than does a gram of clay. This doesn't mean that adding an organic material such as peat moss will automatically increase the CEC by a meaningful amount. The fibrous materials do have some CEC but not nearly as much as the residue left after they have broken down in the soil. This residue represents only a small percentage of the amount of material originally added—it takes a lot of organic matter to make a significant difference to the CEC of a soil.

Depending on where you live, the CEC values in a soil test report may be shown as "milli–equivalents per 100 grams" or "centimoles per kilogram." The numbers associated with these units are exactly the same, so you needn't worry about converting back and forth between two different units. Some reports avoid this issue altogether by not including the units at all. The main thing to remember is the range that represents a low or high CEC value, as shown in the following table.

Typical Cation Exchange Capacity (CEC) of Different Soil Textures and Organic Matter	
Material	CEC (centimoles per kilogram)
Sandy soil	2–10
Loam soil	7–25
Clay soil	20–40
Organic matter (humus)	200–400
Organic soil	25–100

What Else Does the CEC Reveal?

Since the CEC is linked to soil properties such as texture and organic matter, it can be used as an indicator for those properties. A lab can measure the organic matter directly, and when you are standing in the field or the garden, it makes more sense to simply pick up a handful of soil and check the texture by feel.

The lime requirement in some areas is estimated from the soil pH and the CEC rather than the buffer pH measurement. In these cases, the soil test has a lime recommendation, but only measures the soil pH, not both pH and buffer pH. As long as the lab is following calibrations done for your local area, this is a valid way to make the calculations.

Is the Soil Test for the CEC Accurate?

There are ways to measure the CEC that are quite accurate—and expensive. In most cases, rather than measuring the lab calculates the CEC by adding up the cations it measures in the soil. In an acid soil, this corresponds fairly well with the measured values for the CEC. Problems arise, however, when this formula is used with alkaline soils. The soil test extraction dissolves some of the free lime in these soils, inflating the values for calcium and magnesium. When these values are added in with the other cations, the CEC can be significantly overestimated.

What Is Base Saturation?

Soil scientists divide the cations into either "basic" or "acidic." Hydrogen is obviously an acidic cation, as the acidity is determined by measuring the hydrogen ion concentration, but iron and aluminum are also in this group. The basic cations include calcium, magnesium, potassium and sodium. Base saturation refers to the percentage of the CEC that is filled with these basic cations. In a soil with low base saturation, much of the CEC is filled with hydrogen ions and is, therefore, acidic—this is simply an additional indicator of the soil pH.

There has been extensive discussion in some agricultural and gardening publications about the importance of the various basic cations occurring within specific ranges or at specific ratios to one another. In fact, as long as the level of each nutrient in the soil is high enough, there is no advantage in trying to meet specific base saturation percentages or ratios. When a nutrient is truly in short supply, the percentage of that nutrient on the CEC is also low, but the base saturation is not as accurate a way to measure the nutrient supply as is directly measuring the nutrient in the soil.

The one exception to this is sodium. Too much sodium in the soil can harm the soil structure by causing the clays to disperse when they are wet, breaking apart the soil granules into a soupy mess. If the sodium saturation is greater than about 5 percent on a soil test report, watch for structural problems in your soil and consider adding a source of calcium to displace some of that sodium.

Should You Change the Amount of Nutrients Added Because of the CEC or Base Saturation?

For the most part, the answer to this question is no. You can have a small coffee cup and refill it or a large coffee cup that you fill only once, but the end result is that you drink the same amount of coffee. Similarly, just because the soil holds more nutrients does not mean that you have to add more nutrients at one time to give the same supply to the plants.

Of course, every rule has exceptions. In some areas, clay minerals tie up potassium between the clay layers, so you need to add extra potassium to ensure that there is a high enough supply on the outside of the clay, where it is available. Since the CEC is related to the clay content, the recommendations for potassium increase as the CEC increases. Check with your local department of agriculture to see whether this applies to your area, since some labs serving multiple states and provinces may recommend extra potassium on high CEC soils even if it is not required. The extra fertilizer won't hurt the plants, but it is a waste of money and effort.

If you track changes in soil test values over time, you'll notice that the nutrient content of soils with a high CEC changes more slowly than for soils with a low CEC. Just as soils have a buffer capacity for acidity, the CEC represents a buffer against changes (both up and down) in nutrient content. So while you don't need to add any different nutrients to a soil with a low CEC, you should probably test the soil more frequently to monitor for changes in fertility, because these soils are not as well buffered.

How Else Are Nutrients Held in the Soil?

The CEC measures only the capacity to hold nutrients that are cations. What about the rest of the soil nutrients? A number of the essential plant nutrients, including phosphorus and sulfur, are in the form of anions (negatively charged ions), as is the nitrate form of nitrogen. These are obviously not held on the CEC, so how does the soil provide a regular supply of these nutrients?

The small positive charge on the surface of soil particles holds some of the anions, though not many. The majority of anions are either precipitated onto the surface of soil particles as insoluble minerals or integrated into the soil organic matter.

The phosphate ion—the form of phosphorus that plants absorb—combines easily with many elements to create insoluble or slowly soluble compounds. In alkaline soils, the slowly soluble phosphorus is dominated by calcium phosphate that precipitates onto the surface of soil particles. In acid soils, most of the reactions form iron or aluminum phosphates, which are less soluble than the calcium phosphates, so the availability of phosphorus in acid soils is lower than at higher pH values.

The sulfate, or chemical, form of the element sulfur can combine with calcium to precipitate out as calcium sulfate, or gypsum. This is fairly common in semiarid areas, where the soils have a very high calcium content, but it doesn't happen under more humid conditions. Far more sulfur is held in the soil as part of the soil organic matter. Microbes absorb the sulfur to use for their growth, incorporating it into organic compounds. When the microbes die, these compounds slowly break down and release the sulfur back into the soil solution, where it can be utilized by plants.

Nitrogen is part of the soil organic matter as well and is also in the mineral form as ammonium ions that are held on the CEC. At the end of the growing season, nitrogen in the form of nitrate makes up a large part of any excess nitrogen and is easily leached out of the soil. We must ensure that there is a supply of nitrogen for plants each year because, unlike nutrients such as phosphorus or potassium, it does not remain in the soil over the winter.

Soil Salts and Electrical Conductivity

To most of us, salt means one thing: sodium chloride, or common table salt. It's also what the highway department spreads on the roads to melt ice in the winter—and to rust our cars. To a chemist, however, salt is any compound that dissociates into separate cations and anions when it dissolves in water. Along with sodium chloride, then, potassium chloride is a salt, as is potassium sulfate.

This characteristic of salts makes the nutrient portions usable to plants, but it can also lead to problems for plant growth. The concentration of solutes (a dissolved substance) in water is dictated by the number of particles dissolved in that water and not on the mass of those particles. Compounds like sugar dissolve easily in water, but they don't break apart. Each sugar molecule that is dissolved contributes one particle toward the solute concentration. Salt, on the other hand, breaks apart into its separate ions, so each salt molecule contributes two particles to the solute concentration. For example, when sodium chloride dissolves, it breaks apart into sodium and chloride ions, so the concentration of particles is twice as high.

There is a higher solute concentration inside plants than in the soil solution. Plants take in water through osmosis, a process wherein water flows toward the area of higher solute concentration in an effort to equalize the concentrations. This works quite well when the concentration of solutes in the soil solution is relatively low, but imagine what happens when a lot of salts are added to the soil. As the concentration of solutes outside the root increases, the difference in concentration between the inside and the outside of the root decreases, which makes it more difficult for water to flow into the root. At even higher concentrations, water starts to flow the other way, so the soil is actually pulling water out of the root. This is the same process used in making pickles—the high salt concentration in the brine pulls water out of the cucumber.

Think of the impact on plants trying to grow in a salty soil solution. Even though the soil is moist, the plant displays symptoms that look like drought stress because it is unable to pull water into the roots. In severe cases, the roots die back. This is not uncommon when too much fertilizer is placed close to seedlings. When the plants are dug up, the affected roots look burned, as though someone has held a match to them.

Unfortunately, high salt concentration is a common problem in many gardens. There is a misperception that road salt causes most of the salt problems in gardens. While that may be an issue in some scenarios, it is far more often a case of the gardener killing with too much kindness. With concentrated fertilizers, the recommended usage may not look like much when it is scattered on the ground, and it is easy to give in to the temptation to put on a little extra, just to be sure. These extras are on top of the manure and compost that has already been dug into the garden. As this behavior is repeated year after year, the salts can build up to the point where any more fertilizer is harmful for the plants rather than helpful. Organic gardeners should note that it is possible to get fertilizer burn from using organic amendments. The salt concentration of many composts and raw manures is

Salt Tolerance of Various Types of Plants

Salt Tolerance	Perennials, Annuals	Vegetables	Trees, Shrubs
High	Lady's mantle Sea thrift "Karl Foerster" feather reed grass "Burgunder" blanket flower Blue lyme grass Chinese fountain grass Dianthus	Garden beets Asparagus Spinach	Siberian salt tree Sea buckthorn Silver buffaloberry Hawthorn Russian olive American elm Siberian elm Villosa lilac Laurel leaf willow
Moderate	Yarrow "Mönch" Michaelmas daisy Bellflower Garden mums "Moonbeam" thread–leaved tickseed "Fireglow" Griffith's spurge "Elijah Blue" fescue "Bristol Fairy" baby's breath "Stella de Oro" daylily "Palace Purple" coral bells Fragrant plantain lily Evergreen candytuft "Caesar's Brother" Siberian iris Creeping lily turf Eulalia grass	Tomatoes Broccoli Cabbage	Spreading juniper Poplar Ponderosa pine Apple Mountain ash
Moderately low	Hibiscus Geranium Gladiolus Zinnia	Sweet corn Potatoes	Common lilac Siberian crabapple Manitoba maple Viburnum
Low	Primula Gardenia Star jasmine Begonia Rose Azalea Camellia Ivy Magnolia Fuchsia	Carrots Onions Strawberries Peas Beans	Colorado blue spruce Rose Douglas fir Balsam fir Cottonwood Aspen Birch Raspberry
Very low			Black walnut Dogwood Littleleaf linden Winged euonymus Spirea Larch

high enough to damage plants if applied at excessive rates.

To measure the salt concentration in your soil, ask the lab to check the electrical conductivity of a soil–and–water slurry. The more salt there is in the soil solution, the more easily it carries an electric current and the higher the conductivity reading. Ask the lab about the interpretation of the values it provides, because the readings vary based on how the test is done. The sodium concentration in the soil does not tell you whether there is a risk of salt injury, but it may provide a clue as to the source of the salt. High conductivity with low sodium means that there is probably too much fertilizer or manure. High conductivity with high sodium may point to excess road salt.

If your soil has high salt levels, the first step to fixing the problem is to *stop applying more fertilizer*. There are enough nutrients in the soil to meet your plants' needs for many years.

Be sure that the drainage is good and that excess salts have an opportunity to leach out of the soil.

You may have to replace the plants that are sensitive to salt with plants that are more salt tolerant. This may pose problems for the vegetable gardener, since most popular vegetables have a low salt tolerance.

For More Information

- nmsp.cals.cornell.edu/publications/factsheets/factsheet22.pdf
- dpi.nsw.gov.au/agriculture/resources/soils/structure/cec
- gov.mb.ca/agriculture/soilwater/soilmgmt/fsm01s05.html
- ext.colostate.edu/pubs/crops/00503.pdf
- urbanext.illinois.edu/soil/sq_info/saline.pdf
- www1.agric.gov.ab.ca/$department/deptdocs.nsf/all/agdex4246
- agric.wa.gov.au/objtwr/imported_assets/content/lwe/salin/sman/minimisesalinity.pdf

Soil Life and Soil Organic Matter

I find that a real gardener is not a man who
cultivates flowers; he is a man who cultivates
the soil.

— Karel Čapek, *The Gardener's Year*

10

So far, we have addressed the physical and chemical aspects of soil, but now we must delve into the biological nature of soil. In fact, it is the biology that changes bits of broken-down rock and clay into soil. If there is nothing living in it, it is merely dirt; it hasn't yet become soil.

The most obvious sign that soil is home to living organisms is its color. When we look at a dark-colored soil, we immediately assume it is rich and fertile. In many cases, although not always, this is true, which reinforces our attachment to dark soils. Taking advantage of this link in our minds, garden centers make sure that the various topsoils and amendments they sell are dark in color.

What gives soil its dark color? The answer is humus—the complex material left over from centuries of rotting vegetation. Humus forms a significant part of the organic matter in soil. But other parts of the organic matter play different and even more important roles.

The Living, the Dead and the Very Dead

Coined by Fred Magdoff, professor emeritus of plant and soil science at the University of Vermont, this phrase nicely sums up the three fractions, or parts, of the organic matter in soil. The smallest portion comprises the living organisms, from earthworms to fungi and bacteria, that cycle nutrients and create soil structure. But most of the organic matter is material that is in various states of decay: unharvested parts of plants, manure, dead creatures.

Organic Gardening

There is an unfortunate philosophical divide between organic gardeners and those who are not. But every garden relies on the organic materials in the soil and on the living creatures those materials support, regardless of whether we choose to use mineral fertilizers and pesticides.

The claim that harnessing natural cycles can prevent pest and disease infestations and provide all the nutrients a garden needs are occasionally overstated. Yet ignoring these natural cycles inevitably leads to a garden's poor performance.

Bypassing this chapter because you are not an "organic gardener" would be just as foolish as skipping the chapter on nutrients because you don't use fertilizer. Both are important in all systems.

Some is the humus fraction—the material left after many cycles of decay. The living material and the decaying material are sometimes referred to as the active fraction and the stable fraction of organic matter. While it is convenient to divide the organic matter into these two categories, there is actually a continuum from the fresh crop residues to the stable humus fraction.

The Living

Soil is a complex ecosystem teeming with life. A teaspoon of fertile topsoil contains more organisms than there are people on Earth, although most of these organisms are microscopic. This diverse population of organisms performs myriad functions. And while microbiologists have been able to characterize a handful of the groups and species of micro-flora and micro-fauna, they have merely scratched the surface.

We should never underestimate the importance of the creatures that live in our soil, even though we seldom get more than a glimpse of them in action. Without the soil's living organisms to break down the leaves that fall from the trees and the grasses that die back each autumn, we would soon be buried under a mountain of accumulating vegetable matter. What's more, the nutrients in that litter would be perpetually tied up rather than being recycled back to the plants. The soil beneath this layer would never develop the crumbly structure we associate with good topsoil.

The Dead

Fresh organic materials are the fuel that drives the soil ecosystem. Leaves, stems, roots, flowers—all are food for the organisms that live in soil. As plants grow, reproduce and die, they release into the atmosphere some of the carbon in the organic materials of which they are made. They also return carbon to the soil, either through their excreta or as they are eaten. In the process, many nutrients are introduced to the soil and become available to plants.

Organic materials contain bits whose originating plant or animal can be identified, although it may take a microscope to do so. Particulate, or active, organic matter (POM) is sometimes used as an indicator of the health of the soil environment. A lot of POM indicates an ample food source for the soil organisms, which suggests that there is an active soil biology and that the various processes in which these organisms participate—including nutrient cycling and soil structure improvement—are operating well. These materials are not very stable and are broken down and recycled within weeks or months.

The Very Dead

What remains after many cycles of decay is the humus fraction. Nothing is left that can identify where the material originated. Humus contains a lot of carbon, as does any organic material, but it is so tightly bound with other elements that there is very little food value to feed any organism. Fresh organic matter is like meat and potatoes to a bacterium in the soil, while humus is the equivalent of shoe leather. If we had to make a meal of our boots, it would take more energy to break down the leather than our bodies would get back.

While humus is stable in the soil, it is not completely inert, but it breaks down so slowly that its turnover is measured in decades or centuries. The complex organic molecules in humus function as a glue that effectively holds soil granules together, either forming coatings over soil particles or mixing in with clay colloids inside aggregates. Because it is stable, humus does not take part in nutrient cycling, but it has a very high cation exchange capacity, which increases the ability of the soil to hold on to nutrients.

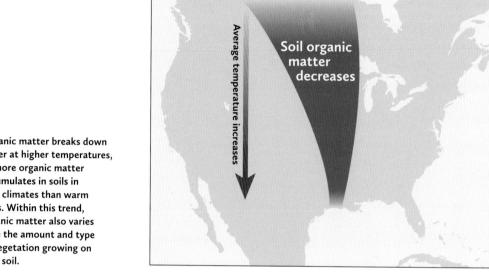

Organic matter breaks down faster at higher temperatures, so more organic matter accumulates in soils in cold climates than warm ones. Within this trend, organic matter also varies with the amount and type of vegetation growing on that soil.

Getting the Right Balance

Provided they are in the right place, all three fractions of organic matter are essential to the proper function of the soil. Adding a bagged soil mix with lots of humus increases the soil's capacity to hold nutrients, because that function is inherent in humus itself. Do not expect an immediate improvement in soil structure, however, because the key factor in achieving the intimate mingling of soil particles is long spans of time.

The amount of organic matter in your soil—and the balance among the three fractions—depends on where you live and how the soil has been managed. There are differences in the natural organic matter content because the combination of organic-matter inputs and decomposition varies from one ecosystem to another. Prairie soil, for instance, tends to be much higher in organic matter than the soil in deciduous and coniferous forests. The fine fibrous roots of prairie grasses add far more organic matter to the soil each year than do the fallen leaves and needles in a forest. At the same time, the moister conditions under the forest canopy encourage a faster breakdown of the organic matter. The net result is higher organic-matter levels under grassland than in a forest.

What Happens to the Organic Materials We Add to the Soil?

Every time you mix pea vines or carrot tops into the soil, you unleash a cascade of biological activity. Insects, mites, snails and earthworms begin tearing the plant material into pieces as they eat their fill, creating residues that smaller organisms can access more easily. Fungal hyphae begin growing through the leaves and stems, excreting enzymes that digest the tough cell walls. Bacteria and other microorganisms colonize the exposed surfaces, absorbing the nutrients that have been released for the plants' growth and activity. All these organisms convert carbohydrates into more organisms, while some is respired as carbon dioxide and returned to the air.

This growing population of fungi, bacteria and other organisms attracts the nematodes and protozoa that graze on this bounty to support their growth. They, in turn, are eaten by other organisms. As these creatures excrete waste products or die, they are cycled through more bacteria and fungi. At each cycle, some of the easily digested organic material is respired and lost, while the most resistant materials gradually accumulate. Eventually, only the toughest material remains—the black substance we know as humus—but it represents just a tiny proportion of what was originally added to the soil. For every 100 pounds (45 kg) of fresh organic material added, a mere 1 to 2 pounds (0.45–1 kg) end up as stable humus.

That's immaterial, however, if you are living in a new subdivision, where the topsoil is routinely stripped off before the houses are built, then replaced with a thin layer of something that looks vaguely dark-colored on top of whatever subsoil remains.

If you're fortunate enough to have a native soil, the humus fraction may still be present in good amounts. It can be lost, however, if there has been erosion by wind or water. Excessive cultivation is often blamed for "burning out" the humus by introducing air into the soil. While cultivation does account for a small loss, most of the losses of the stable organic matter are through its exposure to erosion. Aerating the soil with tillage has more impact on the active organic-matter fraction, since this part breaks down more easily.

The active fraction of soil organic matter comes from regular additions of fresh material. While plants are growing, much of this material is excreted into the soil by the roots. These *root exudates* create a layer of supercharged biological activity in the rhizosphere, the zone around each root. The bacteria and fungi living in this zone help dissolve soil minerals, releasing nutrients in a form that plants can absorb and use. In addition to the

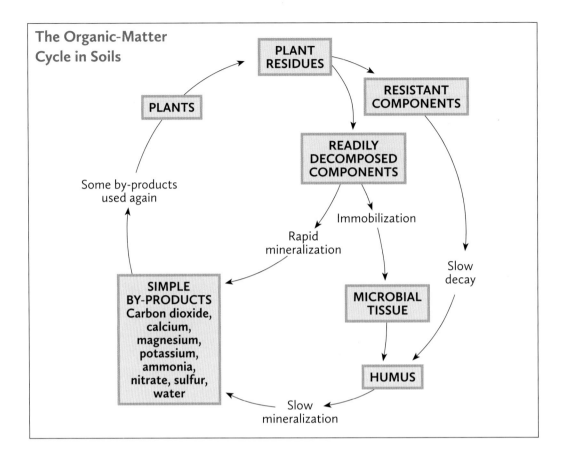

The Organic-Matter Cycle in Soils

PLANTS

PLANT RESIDUES

RESISTANT COMPONENTS

READILY DECOMPOSED COMPONENTS

Some by-products used again

Immobilization

Rapid mineralization

Slow decay

SIMPLE BY-PRODUCTS Carbon dioxide, calcium, magnesium, potassium, ammonia, nitrate, sulfur, water

MICROBIAL TISSUE

HUMUS

Slow mineralization

root exudates, the roots slough off dead cells as they grow and the leaves and stems are often returned to the soil at the end of the growing season.

Gardeners occasionally rationalize cleaning up the unharvested portion of the garden. Some regard dead flower stems and cornstalks as unattractive. Some view the cleanup process as a way to prevent diseases from infecting next year's garden. And some simply want to get rid of stuff that can become wrapped around the rototiller blade or refuses to mix easily into the soil with a shovel or hoe.

The trouble with this habit, however, is that it cuts off the food source for the living organisms in the soil. Eventually, we are left with a soil that hardens like concrete instead of remaining loose and friable. If we don't provide enough food for the organisms in the soil, they start to break down the stable organic fractions just to stay alive, and then we lose the advantages of stable

The Absurd Claims of the Soil-Life Disciples

You may have been told that by achieving the proper balance of life in your soil, you can replace the need for fertilizers—that you can, in essence, create something from nothing.

The reality is that an active soil life unlocks nutrients in the soil, making them more available to plants, but it does so only if those nutrients are present in the first place. Some soil-life techniques appear to work because of the nutrients in the organic materials they contribute and not necessarily because of the organisms themselves. If your soil is deficient in an essential plant nutrient, you must add that nutrient.

The Equally Absurd Claims of the Fertilizer Disciples

Ads on television and in many popular magazines would have you believe that adding a particular brand of fertilizer is exactly what your garden needs to thrive. While it is true that plants require an adequate supply of nutrients to grow well, many garden soils already have lots of nutrients. Adding fertilizer to these soils won't help your plants and may actually hurt them.

soil structure and nutrient retention that the humus provided. Maintaining an active soil biology won't overcome every problem your soil may face, but not having it will guarantee those problems become much worse.

Types of Organisms in the Soil

The usual textbook method of classifying the critters in the soil by species is not useful to most readers. It is more relevant to understand what these organisms do, so here, we have categorized the huge diversity of life in the soil by their functions.

The Shredders

When fresh organic material is added to the soil, the shredders begin breaking it down into smaller pieces. The leaf from a tomato plant may seem pretty small to us, but to a bacterium only a few microns in length, it is huge. To break down a single leaf from the outside is a long, tedious process, even for a large army of bacteria, on the same scale as thousands of humans using their bare hands to move a pile of sand 30 miles (48 km)

Springtail

long, 12 miles (20 km) wide and a mile (1.6 km) deep. As the shredders chew on fresh organic material, they expose surface area that smaller organisms can then access and also start to break down some of the tougher materials.

If you look carefully, you can see some members of the shredder category with the naked eye. Their presence is a good indicator of the overall level of biological activity in your soil. If you don't see any critters when you turn over a spadeful of soil, it's a sign that things below ground are not going well.

The best-known shredders are earthworms. As they burrow through the soil, they eat organic materials that are broken down in their gut, mixed with mucus and excreted. The finely ground material left behind creates a rich buffet for smaller creatures. This process moves organic materials from the surface down into the soil, mixing them together to create topsoil. Evolutionist Charles Darwin spent much of his life studying earthworms, and he estimated that the action of earthworms would completely overturn the soil in a grass meadow in 50 to 100 years.

Many insects fall into the shredder category. Some spend their whole lives underground, while others are in the soil only

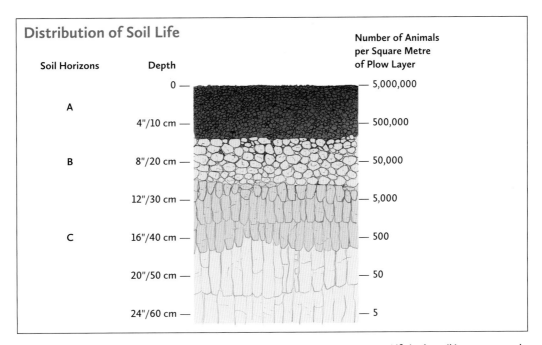

Distribution of Soil Life

Soil Horizons	Depth		Number of Animals per Square Metre of Plow Layer
	0 —		— 5,000,000
A			
	4"/10 cm —		— 500,000
B	8"/20 cm —		— 50,000
	12"/30 cm —		— 5,000
C	16"/40 cm —		— 500
	20"/50 cm —		— 50
	24"/60 cm —		— 5

Life in the soil is concentrated near the surface, supported by an ample supply of food, moisture and air. Creatures inhabiting the deeper layers have adapted to harsher conditions and congregate near roots or earthworm burrows.

during their larval stages. Ants are among the most visible of the shredder insects, and you have probably watched ants carrying cut-up pieces of leaves and stems to their nest. Even more numerous, although harder to see, are the springtails. These tiny, wingless insects (some sources call them proto-insects, "hexapods formerly classified as insects") have an appendage folded under their abdomen that extends when they are threatened, propelling them up to 4 inches (10 cm) into the air. Springtails subsist on decaying vegetation and fungal hyphae in the soil.

Decaying vegetation in the soil is also eaten by many insect larvae, such as the white grub of the June beetle and the European chafer and the wireworm of the click beetle. But when they turn their attention to plant roots rather than dead vegetation, they are officially pests.

Many arthropods make their home alongside the insects in the soil, like centipedes, millipedes and pill bugs, also called sow bugs. Some of them, like wood lice, which are closely related to the pill bug, are specialists, preferring to inhabit rotting wood. There is great diversity among the arthropods, so while many turn to decaying vegetation, others actively hunt for their meals among the other soil fauna.

Earthworms: Nature's Plow

Three distinct groups of earthworms live in our soils and are categorized according to the part of the soil they inhabit. The most visible group is the dew worm, or night crawler (*Lumbricus terrestris*), that makes permanent vertical burrows in the soil. You can usually find the top of the burrow via a midden, a pile of vegetation the dew worm pulls over the entrance to the burrow. The midden acts as food storage and prevents the burrow from drying out.

The second group of earthworms is less visible but more numerous. It includes worms like the "red wiggler" (*Eisenia foetida*) that burrows horizontally through the soil, seldom coming to the surface. Its food source is the organic matter in the soil it ingests as it burrows.

A third group of smaller worms lives exclusively in the litter layer in forests and so has little impact on garden soils.

When conditions are too cold or too dry, earthworms go dormant. Deep-burrowing worms move to the bottom of their burrows, plugging the burrow with their castings as added protection.

This earthworm was not happy that we had disturbed its burrow, but you can clearly see the coating of dark organic matter that has been left behind during its travels. Earthworm burrows aerate the soil and create channels for root growth. Earthworms break down plant residues, mix them with the soil and deposit them deep in the soil.

Shallow-dwelling worms curl up into a tight ball and cover themselves with a mucous cocoon, which keeps them from drying out.

All earthworms breathe through their skin, so they must keep moist, but soil-dwelling worms cannot live for long underwater. Heavy rain that saturates the soil forces the worms to the surface. It creates enough humidity, however, that the worms can travel to new habitats, as long as they are safely under cover before the sun comes out.

Also in this group are the gastropods (from the Latin for "stomach foot"). Snails and slugs belong to this class. While snails spend most of their time on the surface of the soil, slugs, which lack the protection of a shell, prefer to stay under cover to keep their skin from drying out.

Some of the shredders don't wait for plants to die before beginning the shredding process, and at this point, they become pests. When your hostas have been stripped by slugs or the wireworms have nipped off your broccoli transplants, it is hard to appreciate that these creatures are, at least most of the time, more beneficial than harmful. Resist the temptation to wage all-out warfare on the denizens of the soil to protect your plants from the predations of a few. Instead, target your efforts only on the troublemakers.

The Decomposers

In addition to the shredders, fungi play a key role in breaking down big pieces of organic debris. Unlike shredders, however, fungi work from the inside out. Fungal hyphae can grow into decaying leaves, stems and even wood, excreting enzymes that destroy the bonds between cell walls, then digesting and converting lignin and cellulose into simple sugars the fungi can use. In the process, the strands help bind the soil particles together and build a stable soil structure.

Most fungi in the soil are microscopic, but some species form visible networks of hyphae, particularly on rotting wood. The fungal parts that are most frequently seen are the fruiting bodies. Mushrooms and toadstools that pop out of the soil after a warm rain are only a tiny part of the fungal network below the surface.

The Digesters

Once the organic matter has been broken down into smaller pieces, bacteria and actinomycetes go to work. Through their sheer numbers, these organisms are able to access most of the easily digested materials in the soil and incorporate them into their bodies, with the sole purpose of making more bacteria. In the process, they release nitrogen, phosphorus, sulfur and other

Actinomycetes

Actinomycetes are an odd class of organism. Bigger and slower-growing than bacteria, they are less aggressive than fungi at colonizing decaying materials. After a cursory look, you might expect these "thread bacteria" to be pushed to the furthest corners of the soil environment, unable to compete with their tougher and faster neighbors. In fact, actinomycetes thrive in most soils around the world. That fresh, earthy smell gardeners associate with a warm spring rain is actually geosmin, a compound released by actinomycetes.

The secret to their success? Chemical warfare. Actinomycetes, along with a few fungus species, produce compounds that kill bacteria. These antibiotics allow actinomycetes to get at food sources that would otherwise be gobbled up by other creatures. We take advantage of this compound whenever we get a prescription to treat strep throat or some other infection. Any antibiotic with "mycin" in its name has been isolated from actinomycetes, the same way that penicillin is extracted from the penicillium bread mold, which also inhabits the soil.

nutrients that have been bound up in the organic matter. Some of the nutrients are used by the bacteria and actinomycetes, and some are released into the soil solution, where they are available for uptake by plants.

Microbiologists have identified only a small number of the bacterium species in the soil. There is an incredible diversity, adapted to a wide range of soil conditions. Some need aerobic environments to survive, while others grow only in the absence of air. Different species have different requirements for temperature, pH, food supply and moisture. In most cases, the key to attracting a desirable population of microbes is not to add inoculants to the soil but to provide the proper environment. When conditions are right, there is usually enough of a reservoir of cells to multiply and populate the soil. If conditions aren't right, the inoculated cells won't survive anyway.

The Grazers

Bacterial and fungal growth attracts a whole population of tiny animals that feed on them in much the same way that cattle or sheep graze a pasture. These include protozoa, such as the amoeba and paramecium. (In high school, you may have looked through a microscope at these creatures swimming around in droplets of pond water; their relatives live in the water films in soil.) Large numbers of mites and nematodes also fill this role. They can be highly specialized—some nematodes feed only on fungi, while others seek out bacteria.

Not all the nutrients consumed by the grazers are used for their own growth, and the waste they release hastens the cycling of nutrients into a form that plants can absorb.

The Hunters

Just as in aboveground ecosystems, there are predators in the soil. Some predation is accidental—an earthworm might ingest an amoeba along with the rotting vegetation it is feeding on—but more often, predators are specialized hunters. This group includes many species of nematodes, mites and small insects. Aside from keeping the population of grazing animals in check, these predators continue the cycling of nutrients through the soil ecosystem.

In a strange twist, there are even predatory fungi. One particular species has developed specialized hyphae that form a noose-like structure just large enough for a nematode to swim through. When the nematode enters, the noose tightens around its body, ensnaring it. The fungal hyphae then grow into the nematode, digesting it from the inside out.

The Fixers

One group of microbes plays a crucial role in the soil environment by taking nitrogen out of the air and "fixing" it in a form plants can use. Without this process, natural ecosystems would soon run out of available nitrogen to support plant growth. Most nitrogen fixation is carried out by bacteria that live symbiotically with legumes, but a few species of bacteria and blue-green algae fix nitrogen without being associated with higher plants.

The Symbionts

When two organisms live together for their mutual benefit, it is called symbiosis, and each of the organisms is a symbiont. One example of a symbiotic relationship is that of rhizobia and the roots of legumes. Rhizobia attach to the root hairs of a host plant, which responds by growing around the bacteria

Nematodes

Nematodes that live in the soil are tiny, active worms, typically less than ¹⁄₂₅ inch (1 mm) in length. They have gotten a bad rap in agricultural settings because a few species are plant parasites. But they represent only a small minority of all the soil-dwelling nematode species. Nematodes can be divided into the following groups:

Grazers eat fungi and bacteria. The fungi feeders have a narrow stylet, or appendage, that they use to puncture the fungal hyphae and suck out the insides. The bacteria feeders have liplike structures for slurping up small bacteria.

Hunters search out other nematodes, protozoa, such as paramecia and amoebas, and even small mites and insects.

Plant parasites feed on plant roots. These nematodes are pests to farmers and gardeners alike. They have a longer, thicker stylet than the fungi feeders and are able to penetrate the cell walls of the roots. Some species burrow into the roots and form cysts.

Insect parasites burrow into soil-dwelling insect larvae and kill them before moving on to look for new hosts. You may have bought nematodes at the garden center that are parasitic on white grubs. These can provide very effective control for insects that would otherwise damage our plants.

to form a nodule. The bacteria in this nodule get food from the plant, which they use to fix nitrogen from the air in a form the plant can use. Both the plant and the bacteria benefit from this relationship. Common legumes include beans, peas, clovers and alfalfa, along with a few tree species, such as locusts. The rhizobia species involved in this symbiosis are specific to their host plant, so the bacteria that form nodules on clover roots are different from the bacteria on pea roots. This is one instance where adding a bacterial inoculant can be beneficial. If you are introducing a new legume to your garden, there won't be a population of the correct rhizobia. Just be sure the inoculant you buy is right for the plant you are growing.

Another symbiotic relationship that covers a much broader range of plants occurs between mycorrhizal fungi ("fungus root") and plant roots. These fungi form networks that penetrate plant roots at one end and extend out into the soil at the other. Far from being harmful, mycorrhizal fungi pay back the small amount of carbohydrates they take from the root by acting as an extension of the root system. Plants with symbiotic mycorrhizae can extract more water and nutrients from the soil than plants that haven't been colonized by these fungi. Mycorrhizae

Ants

Ants are among the more visible inhabitants of the soil. While they live underground, they spend a good part of their lives aboveground foraging for food. When this foraging brings them into our houses, they become pests. But should we worry about them in the garden?

For gardeners, the benefit of having ants around outweighs any harm. Their tunnels aerate the soil and improve drainage. They mix organic materials from the surface deep into the soil as they carry food items to their nests. And, as part of their nest building, they glue soil particles together into a very stable structure. Of course, we can

sometimes have too much of a good thing. I don't object to leveling the mounds that an active ant colony can build up, even to the point of using a commercial ant killer to

knock back the population. And if you live far enough south that you have fire ants, you can avoid some really nasty stings by eliminating that particular species from the garden.

are particularly valuable for extracting phosphorus from the soil, since they release chemicals that help dissolve some of the phosphorus compounds not normally available to plants.

Most common garden plants are colonized by mycorrhizae. This symbiosis does not appear to be as species-specific as the relationship between rhizobia and legumes. The largest exception is the brassica, or mustard, family, whose members do not support mycorrhizae. It is possible to have low levels of mycorrhizal fungi in the soil after growing a crop of broccoli or cabbages. Very occasionally, plants that follow brassica crops may suffer from a phosphorus deficiency early in the season because the symbiotic infection doesn't occur quickly enough. However, this happens only in soil that is already low in phosphorus.

You can encourage a healthy mycorrhizal population by minimizing how often and how intensely you cultivate. In undisturbed soils, the network of fungal hyphae is well established and ready to colonize new roots soon after they emerge. Tillage, however, breaks up these networks and kills many of the mycorrhizae. Bare soil doesn't provide a food source for the symbiotic fungi, so planting cover crops like rye or clover also helps build mycorrhizal populations. This is especially helpful after you've harvested a brassica crop.

The Pathogens

Most of the organisms in soil are beneficial—but not all. Unfortunately, the "bad actors" give soil life a bad name. They can be bacteria (e.g., bacterial wilt of beans), fungi (e.g., verticillium wilt, Phytophthora root rot), nematodes (e.g., root lesion nematodes, dagger nematodes) or insects (e.g., white grubs, wireworms). A healthy, diverse population of soil organisms helps keep these pathogens under control but does not eliminate them completely.

Some organisms living in the soil can make us sick if we don't take reasonable precautions. Human skin is a very effective barrier against infection, and most of the bacteria in the soil are not very infective to people—our immune systems can easily handle the few that do get through. The exception, however, is when soil gets into a deep puncture wound. Clostridial bacteria spores are present everywhere, but these bacteria can grow only in the absence of air. A deep puncture wound provides an ideal environment for them to grow and multiply. They then release toxins that can cause maladies such as tetanus. It is important that your tetanus shots are up to date, and if you happen to get a puncture wound while working in the garden, follow up with your doctor.

Is There Value in Mycorrhizal Inoculants?

The positive link between plant growth and mycorrhizae is well established, but this does not necessarily mean that inoculating the soil with mycorrhizae will help your plants grow better. Trials with these inoculants have been inconsistent. The conditions at each site should determine whether the inoculant will benefit the plants. Inoculants may be helpful where there isn't a viable population of mycorrhizae, which is often the case following a flood or in a new subdivision, where a thin layer of topsoil has been laid down over highly disturbed subsoil.

Inoculants may still fail under these conditions. If the inoculant has not been properly prepared, handled and stored, there may be no viable spores left to colonize the plant roots. Even viable inoculants can fail if they are not adapted to the soil conditions at the site. Although the symbiosis is not species-specific, there is good evidence that it may be soil-specific.

For More Information

- The USDA-NRCS website has an excellent introduction to soil biology at **soils.usda.gov/sqi/concepts/soil_biology/biology.html**
- *Teaming with Microbes: The Organic Gardener's Guide to the Soil Food Web* (revised edition) by Jeff Lowenfels and Wayne Lewis. Portland, Oregon: Timber Press, 2010. While the authors may exaggerate their claims of what microbes can do, this text is a good introduction to soil biota, and the photos are excellent.

Harnessing Biological Cycles for Our Benefit

One is constantly reminded of the infinite lavishness and fertility of Nature—inexhaustible abundance amid what seems enormous waste.

— John Muir, *My First Summer in the Sierra*

11

THE SOIL IS an integral part of all terrestrial cycles, and a key point to understanding any cycle is to remember that while nothing is created or destroyed, any element can change forms. At each step in the cycle, we try to understand where the carbon, nitrogen and sulfur go, since the total amount must balance out in the end. Once we understand this concept, we can start to harness these cycles for our benefit.

The Carbon Cycle

Although we discussed some aspects of the carbon cycle and organic-matter breakdown in the previous chapter, we didn't address the role of photosynthesis, a process that allows plants to convert carbon dioxide from the atmosphere into simple sugars. Photosynthesis is the critical piece in the creation of all the organic materials we eat and those returned to the soil as unharvested plant parts.

Once carbon has accumulated in plants, it may pass through several bodies as it decomposes. At each step in this process, the organisms that extract energy from the carbon compounds respire carbon dioxide, returning a large part of the carbon to the atmosphere, where it is available to be absorbed by plants again. A residue of complex carbon compounds is left behind. While these compounds are tough to break down, all that eventually remains of the original carbon input is a small amount of humus mixed with soil minerals.

Organic carbon accumulates in soil when more is added than is lost each year through such processes as respiration and soil erosion. This is how rich, dark topsoil is created. Unfortunately, much of what we do to soil increases the rate of carbon loss, which hurts its ability to hold water and nutrients. It also affects how well its structure stands up to heavy rain. If a healthy garden is our goal, this trend must be reversed. Later, we'll explore ways of building soil organic matter, but first, let's look at the link between carbon and nitrogen in soil.

Carbon:Nitrogen Ratios of Common Organic Materials	
Bacteria	5:1 to 7:1
Fungi	9:1 to 20:1
Soil organic matter	10:1 to 12:1
Grass clippings	17:1
Raw manure	20:1
Composted manure	10:1 to 20:1
Yard-waste compost	15:1 to 30:1
Peat moss	60:1
Leaves	60:1
Cereal straw	80:1
Sawdust	200:1 to 500:1
Corrugated cardboard	600:1

The Dance Between Carbon and Nitrogen

Along with carbon (C), organisms in the soil also contain nitrogen (N), a key part of protein. In bacteria, the proportion of these two elements is five to seven parts carbon to one part nitrogen. As bacteria feed on an organic substrate (like leaves and stems returned to the soil), they absorb both carbon and nitrogen. Extra carbon is necessary to support respiration (i.e., to be lost as carbon dioxide), but as the bacteria grow and multiply, they maintain an internal C:N ratio of about 5:1.

If there is more nitrogen in the substrate than the bacteria need, they release it back into the soil solution as ammonium. If there is not enough nitrogen, the bacteria pull nitrogen from the soil solution; otherwise, the amount of nitrogen available would limit the bacteria's growth. (The same process occurs with fungi, but the internal C:N ratio of fungi is higher, between 9:1 and 20:1.)

This ratio means the proportions of carbon and nitrogen in added organic materials greatly influence how quickly the materials break down and whether nitrogen will be available to plants right away or later in the season. The C:N ratios of common organic materials are shown in the table at left.

When organic material is added to soil, the population of bacteria and fungi rapidly increases to take advantage of the new food source. If the material has a low C:N ratio, there is more nitrogen than the microbes need, so mineral nitrogen is released into the soil solution, where plants can absorb it. The rate of microbial growth is limited only by the amount of carbon that is available.

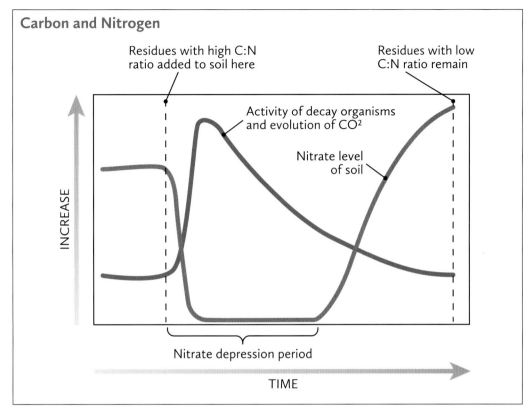

Carbon and Nitrogen

Residues with high C:N ratio added to soil here

Residues with low C:N ratio remain

Activity of decay organisms and evolution of CO_2

Nitrate level of soil

Nitrate depression period

INCREASE

TIME

The picture is quite different when organic material with a high C:N ratio is added to soil. In this scenario, there is an abundance of carbon for the bacteria and fungi to eat but not enough nitrogen. Remember that there are a *lot* of bacteria and fungi in soil, so they quickly start to grab nitrogen out of the soil solution. When the nitrogen is tied up in microbial bodies, it is not available to plants. The feeding frenzy continues until the supply of carbon compounds is used up by microbial respiration. Once the food supply is gone, the microbial population crashes, and the nitrogen that has been tied up is released back into the soil solution. When organic material high in carbon is added, nitrogen eventually becomes available again, but there can be a significant period during which plants suffer from a shortage of available nitrogen.

Does applying bark mulch tie up soil nitrogen? Bark mulch has a high C:N ratio, and if it were mixed into the soil, it could tie up nitrogen, but when the mulch is applied to the surface,

Adding materials with a high C:N ratio will temporarily deplete the available nitrogen in the soil.

there is not enough contact between it and the soil to cause any serious problems. Because the mulch decomposes slowly, it does not create the rapid flush of microbial growth that ties up soil nitrogen.

The exact ranges of C:N ratios considered "high" and "low" depend on the size and type of material. As a general rule, if the C:N ratio is less than 20:1, there is a net release of nitrogen into the soil solution. If the C:N ratio is greater than 40:1, expect the nitrogen to be tied up. In between these values, the soil nitrogen may change very little or may go up or down. If the carbon is easily digested, soil nitrogen is suppressed, because the decay organisms grow faster. If the carbon compounds are difficult to break down, either because they are chemically resistant or because they are in big chunks, there is less suppression of soil nitrogen, and that suppression won't last as long. Weather conditions also have a bearing. Microbial activity is faster in warm, moist conditions, so the suppression of available nitrogen from high C:N materials is even more severe with warm weather, but the nitrogen supply in the soil returns to normal fairly quickly.

To enable the C:N ratio to work favorably, add high-carbon materials in late summer, after the crops are harvested. The soil nitrogen suppression then occurs when garden plants don't need nitrogen, and the addition helps to hold some nitrogen that might otherwise be lost over winter. Mix these materials into the soil around perennial plants that are hardening off for the winter. Remember that too much nitrogen from the soil encourages tender growth, which might freeze over the winter. To make nitrogen available to feed plants in spring, use materials with a low C:N ratio.

The Nitrogen Cycle

Biological cycles affect nitrogen more than any other plant nutrient. And that creates headaches when it comes to measuring the amount of available nitrogen in the soil, since nitrogen never stays in the same place or the same form for long. But biological activity affords opportunities to manage nitrogen so that a supply is available to plants for the entire season.

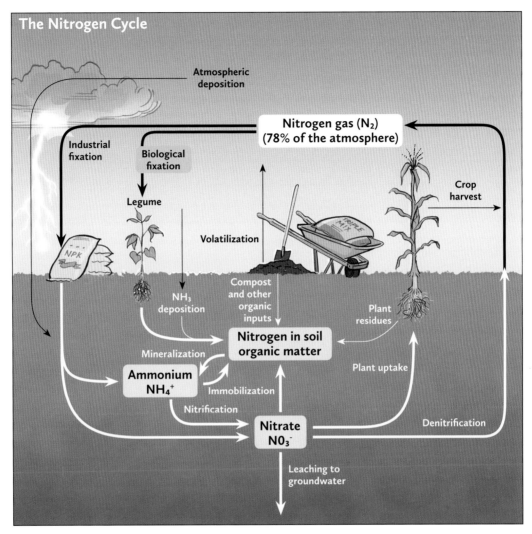

Nitrogen is continually cycling through different forms in the soil, the plants and the atmosphere. This makes it challenging to measure and to manage.

The air we breathe is 78 percent nitrogen, but that nitrogen is in a form plants can't use. The largest reservoir of nitrogen in the soil is bound in soil organic matter, which is also unavailable to plants. The biological activity in the soil transforms nitrogen gas or organic nitrogen into forms that plants can use.

Fixing nitrogen out of the air is accomplished by a few specialized bacteria or blue-green algae living in the soil either free or symbiotically with legumes. Operating at the soil's ambient temperature, these little creatures accomplish what requires high heat and pressure in the industrial process that produces nitrogen fertilizer. The only other natural phenomenon that can convert nitrogen gas to a nitrate form is the intense electrical charge of a lightning bolt, which provides only a tiny part of the total nitrogen in the soil.

Microbial action breaks down organic matter in the soil and releases some of the nitrogen as ammonium—the same form of nitrogen produced by bacterial nitrogen fixation (and in most nitrogen fertilizers). Some of this ammonium is taken up by other microbes, particularly if there is lots of carbon in the soil that can serve as food for the microbes. Some of the ammonium, however, is converted to nitrate by bacteria in a two-step process. While plants can use ammonium, they absorb most of the nitrogen they take up in the nitrate form, because it moves easily with the water flowing toward the roots.

Nitrate that is not taken up by plants can be lost when it leaches deep into the soil or when the soil becomes saturated and the nitrate is converted back to nitrogen gas in a process called denitrification. Bacteria also carry out this activity by using the oxygen from the nitrate molecule for respiration if free oxygen isn't available.

The nitrogen taken up by plants is eventually returned to the soil with the unharvested portion of the plant. On a livestock farm, part of the harvested nitrogen is returned to the soil in livestock manure. This recycled plant material becomes part of the organic nitrogen pool and is cycled back into the available nitrogen for plant uptake.

The Sulfur Cycle

Microbial action causes sulfur to cycle through a number of forms. While the sulfur pool in organic matter is not as large as the nitrogen supply, it nevertheless serves as an important reservoir. Sulfate sulfur, the form plants use, is released during organic-matter decomposition. It can be taken up by microbes in the soil and returned to the organic pool. There can even be gaseous losses of sulfur from the soil, although this occurs only in soils that are flooded for extended periods. When the bacteria in the saturated soil have used up all the oxygen gas and the oxygen on the nitrate molecules, some species of bacteria start to pull oxygen off the sulfate molecule. The end product of this process is hydrogen sulfide. You may get a whiff of its characteristic rotten-egg smell when you dig in a really wet corner of your yard.

Other Nutrient Cycles

The essential plant nutrients cycle through organic materials, but none of them builds up as extensive a reserve in the organic fraction as do nitrogen and sulfur, and they lack the diversity of forms that characterize nitrogen and sulfur. For most nutrients, the chemical bonds to the clay minerals or to the cation exchange capacity of the humus in the soil are far more important than the transformations by soil microbes.

Ensuring an Active Soil Biology

Keeping the life in soil happy and healthy doesn't solve all problems. It can't magically create nutrients where there aren't any, and it won't overcome the adverse effects of severe compaction. What it does do is ensure that any chemical or physical management of the soil works better. In addition, the conditions that are good for the biology of our soil are also good for our plants.

The basic needs of the life in soil are not much different than yours or mine: food, water, air, a secure home, temperatures that aren't too hot or too cold and an absence of toxic substances. There are huge differences, of course, in what these necessities mean to us as compared with their effect on a bacterium or a nematode.

The one factor that isn't included in this list of basic needs is the soil pH. Microorganisms in the soil have a definite preference for the pH range of their environment. Many of the processes bacteria perform in soil are diminished or shut down completely when the soil pH is too low. You can see this when you try to grow a legume, such as beans, in an acidic soil. The beans are deficient in nitrogen because the bacteria in the nodules are not working. Fortunately, fungi are not as susceptible to acidic soils, so biological activity does not stop completely, but it does slow down significantly. Applying lime to sweeten an acid soil helps the plants as well as other organisms.

A steady supply of fresh organic material ensures that the soil microbes are well fed. When plants are growing, exudates from the roots provide this nutrition, and at the end of the season, you can make a big difference to the life in your garden by properly managing the plant leftovers. Cleaning up all the unharvested parts of plants and shipping them off to the municipal compost facility may be a good disease-control strategy, but it cuts off the food supply for the soil life. The organisms living in the soil can't travel to find new food sources. Their options? To begin eating whatever organic matter is left in the soil, to go dormant or to die. None of these lead to better soil structure or nutrient recycling.

Keep soil biology active by adding fresh organic matter to the system or by planting a cover crop to generate a new batch of roots to feed the soil biology. Organic matter should be a mix of easily digested green materials (like lawn clippings or manure) and more resistant materials (like fallen leaves, mature plant stems or straw) that keep the activity going longer. Just as we can't live on sugar alone, life in the soil benefits from a bit of protein and fiber.

Biological activity in soil takes place in the water films that surround soil particles, so dry soils significantly slow the activity of all types of soil microbes. Some fluctuations in soil moisture and biological activity are to be expected, but in the absence of water, reduced biological activity can suppress the availability of nutrients to plants. Even after a rain, there is a time lapse before the nutrient supply kicks in. A layer of mulch not only evens out these moisture fluctuations but provides a continuous food source to the microbes near the surface.

Humans need oxygen to breathe, and so do most microbes. Some can switch from oxygen gas to pulling oxygen from compounds in the soil. The first step in this process is denitrification, since nitrate is one of the easiest compounds to break down. After the nitrate is used, bacteria can pull oxygen from sulfate, iron oxide and even sugar (releasing methane and water instead of carbon dioxide). However, the conditions that allow these reactions to occur are not healthy for the plants themselves, so even if soil microorganisms are able to adapt to a lack of oxygen, it is better to keep their environment aerated.

A secure home for a microorganism is one that isn't continually being disrupted by external forces. This is especially true for fungi and actinomycetes, whose hyphae (the fungal equivalent of roots) can extend for considerable distances. Frequent, aggressive cultivation upsets the balance of biological activity in your soil, favoring bacteria species that multiply rapidly. Unfortunately, these species may not do the best job of building soil structure or cycling nutrients. Try not to cultivate more often or more deeply than is necessary.

Soil temperature greatly influences how active soil organisms are. As the soil temperature rises to roughly 85°F (30°C), everything moves faster. Once that temperature is reached, however, activity drops off quickly. Fortunately, soil moderates temperature swings, particularly farther below the surface. You can help this process along by keeping plants growing in the garden, since they protect soil from direct sunlight, and by having a layer of mulch in place. Mulch keeps soil warm when the air temperature drops in the autumn, but the trade-off is that it also slows the warming of soil in the spring. Although we assume that biological activity stops when the soil temperature drops to the freezing point, this is not quite true. Some respiration is still going on, but it is very slow. The natural organic-matter content of soil increases as you move from the equator toward the poles, because there is less decomposition during the winter months.

The final way to ensure an active soil biology is to keep toxic substances out of the soil. Because of the diversity of life in the soil, toxicity to a microbe is not the same as toxicity to a human. Microbes quite happily eat substances that are poisonous to us, as long as the concentrations are not too high. Soil organisms are adaptable enough that toxicity is not a problem as often as is starvation. What *is* harmful to life in the soil is high concentrations of salts and heavy metals. To keep everything happy below ground, go easy with fertilizers (organic and inorganic) and don't use chipped pressure-treated lumber for mulch.

Building Soil Organic Matter

Improving the physical condition of soil, building the diversity of life that lives in it and increasing the effectiveness of the nutrient cycles in meeting plant needs all depend on adequate organic matter. This is a balance between the quantity of organic matter that is added each year and the quantity that is lost through decomposition and erosion. If we remove all the unharvested parts of plants during fall cleanup without adding any other organic amendments, the soil rapidly loses organic matter. The ground becomes harder and more subject to crusting after a rain, and plant growth declines.

It requires a conscious effort to reverse this trend. Organic material must be added in such a way that it becomes part of the soil rather than just another layer on top. In fact, improper additions can create more problems than they solve.

The best way to build organic matter is to grow more plants and return them to the soil. Rotating plantings so that part of the garden is planted with a perennial grass crop builds up organic matter with the root exudates and the tops of the plants, which are worked back into the soil. Even better is a mix of grasses and legumes—a tall fescue and red clover mix, for instance—that adds nitrogen as well as organic matter to the soil. The grasses should be left to grow for the entire year, with the tops periodically cut and left on the surface to rot down.

Make these strategies part of a program in which no plant is grown in the same soil for two years in a row. Rotating crops to different parts of the garden effectively breaks insect and disease cycles and helps improve root growth. Plant roots will follow the old root channels of other plants but will avoid the channels left behind by the same species of plant. Rotation also allows plants that don't need a lot of nutrients to scavenge what is left over from the high-demand plants. If you are going to manage your garden organically, crop rotation is absolutely essential and is beneficial for gardens that receive fertilizer as well.

There are several methods of rotating crops in the garden; the chart on page 193 illustrates one way to group plants. In this system, you divide the garden into six sections. The plants in

Group 1 are planted after the forage crop (Group 6). The next year, the Group 2 plants replace the Group 1 plants, and so on. You don't have to follow this order precisely, but it is important to avoid planting members of the same group in the same plot for two seasons in a row.

Crop Rotations in the Vegetable Garden

Group 1: Leaf Crops	Group 2: Legumes	Group 3: Solanaceae	Group 4: Root Crops	Group 5: Curcurbits	Group 6: Forages
Sweet corn	Beans	Tomatoes	Carrots	Pumpkins	Ryegrass
Brassicas	Peas	Peppers	Beets	Squash	Tall fescue
Lettuce	Snow peas	Eggplant	Onions/Garlic	Melons	Red clover
Spinach		Potatoes	Radishes	Cucumbers	Alfalfa

Not everyone has the space or the inclination to leave part of the garden planted with a nonproductive crop for an entire year. Many of the same benefits can be realized by planting cover crops or green manures when there is bare ground after a crop is harvested. Alternatively, you can add cover crops to a rotation that already includes forages to speed organic matter accumulation. Some plants are sold specifically as cover crops, such as tillage radish, but you can also plant leftover vegetable seeds or cereal grains, like barley, oats or rye. Cereal crops are divided into spring and winter varieties, each of which offers advantages. Spring varieties keep growing in cool weather, adding more growth in the fall, then dying when there is a hard freeze. Winter varieties tend to go dormant with the cool fall weather, growing roots instead of tops, but the tops grow quickly in the spring. Be sure to turn winter cereals into the soil when they are about knee-high, or the straw will become too tough to break down in time to plant vegetable seeds.

Cover crops build soil structure, break up shallow, compacted layers and suppress weed growth, but they need lots of nitrogen to grow well (except for the legumes, like peas or clover). If manure or fertilizer is not applied when cover crops are planted, or if there is not a lot of nitrogen left in the soil from over-fertilizing the previous crop, they are probably not worth the effort and expense.

Organic matter from weeds is just as useful as organic matter from other plants, so why don't we just let the weeds grow? The trouble with weeds is that they don't know when to quit. To survive, they have developed coping mechanisms. Weeds produce a lot of seeds early in their growth. They also tend to keep germinating over a long period, so rather than emerging all at once, they come up in successive waves. If it were possible to kill off the weeds before they produced any seed, weed growth would, indeed, offer a benefit without a risk to the next crop. Unfortunately, we'd have to kill off the weeds before they were large enough to generate a meaningful amount of organic matter.

Organic Amendments

Despite the value of cover crops and forage, many gardeners prefer to add organic amendments instead. One reason is that it is difficult to establish a crop rotation in a perennial flowerbed. And it's often necessary to add an amendment simply to bring the garden to a state where anything will grow.

A wide variety of materials will add organic matter to the soil, but most garden centers seem to offer only peat moss or composted manure. Gardeners in rural areas may have access to various kinds of raw manure, hay or straw. Many urban communities make yard-waste compost available to their residents, while home gardeners often maintain a compost pile in their backyards. The chart opposite outlines the pros and cons of a range of organic amendments.

Remember that organic materials contain many of the nutrients plants require. The problems appear when too many nutrients are applied. They then change from beneficial nutrients to harmful salts. The advantage to an amendment like peat moss is that it doesn't contain much except cellulose and lignin, so it is almost impossible to overapply. Of course, the drawback is that peat moss can't do anything to overcome a shortage of nutrients in the soil. (You'll find extensive discussions about the sustainability of amending soil with an organic material mined from wetlands on the Internet.)

Compost

Material	Characteristics	Advantages	Challenges
Peat moss	Fine, fibrous, dry	Weed-free, odorless, consistent	Doesn't provide much usable food for soil organisms
Straw	Coarse, fibrous, dry	Fairly consistent, easy to handle	Can be difficult to mix with the soil; may tie up nitrogen; may have weed seeds
Hay	Mix of fine and coarse, dry	Legume hay provides nitrogen and organic matter	Can be difficult to mix with soil; may have weed seeds
Sawdust	Fine, granular, dry	Weed-free, consistent	Ties up nitrogen in soil; doesn't provide much usable food for soil organisms
Raw manure: Cattle Sheep Chicken	Mixture of fine, coarse and wet materials; inconsistent	Good source of nutrients and organic matter; hen manure also contains lime	Can be difficult to mix evenly with soil; smelly; may have weed seeds; too much may cause salt injury to plants
Composted manure	Fine, moist, somewhat sticky	Relatively easy to handle, consistent, easy to mix with soil	Nutrients are only available slowly; some are high in salts
Municipal yard-waste compost	Fine, friable	Relatively easy to handle, consistent, easy to mix with soil	Relatively low concentration of most nutrients; not a lot of food for soil organisms
Homemade compost	Variable	Should be the same as municipal compost, but it depends on who's making the compost	Variability: ranges from wonderful to awful

Why Compost?

I am a bit of a skeptic about compost and the suggestion that it is a magical solution to all our garden problems. When compost is promoted as a cure-all, it becomes difficult to sort through the hype to identify its pros and cons.

There are lots of advantages to composting. The materials produced through the composting process are much easier to handle and mix with soil, since the big chunks are broken down. Unpleasant odors are almost eliminated. Weed seeds and plant diseases are killed (provided the compost pile is hot enough). Nutrients are stabilized in forms that are slowly released in the soil.

Other advantages attributed to compost are less certain. Is the mix of aerobic microorganisms in compost better for the soil than the microorganisms in uncomposted materials? I don't know for certain, but I am pretty sure that the population of

How Much Organic Matter Is Too Much?

It takes a lot of added organic matter to build up a depleted soil because so much of the material rots away before turning into humus. What happens, then, if we add a 6-inch (15 cm) layer of organic material all at once?

The first challenge is mixing the organic material with the soil. Even with a rototiller, it takes several passes to mix it adequately, and the organic material is still not fully combined. It takes biological activity to break down the organic matter and mix it intimately with the soil's mineral particles to form stable granules. While that time-consuming process is taking place, nitrogen is either tied up in the soil or released into the soil. All this organic material can immobilize some of the micronutrients or change the pH of the soil. And as the material decomposes, it may release compounds that are harmful to plant growth.

The volume of material rapidly declines as the organic material breaks down and is incorporated into the soil, and the surface of the soil subsides. Recent transplants may become stranded above the surface or sink into a hole, and neither scenario is desirable.

A working rule of thumb is to add no more than about 10 percent of the total soil volume in a garden at any one time. Adding reasonable amounts of organic materials over a number of years produces far superior results than trying to do it all at once.

microorganisms in the soil will overwhelm the critters in the added material within a day or two, unless the application rates are ridiculously high. Does composting retain nutrients? A lot depends on the process. There is evidence that even with a well-managed composter, a significant fraction of the nitrogen in the raw materials can disappear during the composting process, lost to the air as ammonia.

One of the key debates is whether compost is better for improving soil structure than adding raw organic materials. Logic dictates that since it is the fungal hyphae and bacterial slimes produced during decomposition that help create a stable soil structure, there is a greater benefit to having decomposition occur in the soil. With composting, most of the biological activity happens outside the soil environment. But if the choice is between composting your old pea vines and returning them to the soil or leaving them on the curb for the garbage truck, I vote for composting!

Compost bins can take many forms, but they all perform the same function—breaking down organic materials to recycle back to the garden.

How to Compost Successfully

Composting is essentially accelerated rotting. Many products created to assist in composting do less for the quality of the finished compost than they do for simply speeding up the process. There are advantages to the rapid breakdown of the material added to the compost pile—but only up to a point.

The key to composting is to create the right environment for bacteria and fungi to do their work, which means having the proper mix of air, water, carbon and nitrogen. The composting process is aerobic (taking place in the presence of oxygen), so the pile either should contain coarse materials to keep it loose enough to allow air to flow through it constantly or should be turned over regularly to expose the materials to the air. Microorganisms need water, so don't allow the pile to dry out. If there is too much water, however, it's difficult to keep the pile aerobic. Carbon compounds, as a general rule, are derived from materials that are dry and brown, such as wood chips, leaves, bark and clippings from mature plants. Nitrogen comes from materials that are green, like lawn clippings and vegetable waste.

But every rule has its exception, so while manure may appear brown, it falls into the green category, because it contains lots of nitrogen.

Starting a compost pile is relatively simple. Alternate layers of green and brown materials so that there is about twice as much brown as there is green. Make sure it is moist but not too wet, then let the microbes do the work. The art to composting is to make sure air can get through to the center of the pile to keep the process aerobic. There must be a big enough mass of material to hold the heat and moisture in but not so big that the center goes anaerobic.

To help air get into the compost pile, make sure there is a good supply of coarse material mixed in with the finer material. Since we usually want the compost to come out in small pieces that will mix easily into the soil, this may seem counterintuitive, but it is necessary. Coarse stalks from the garden work well, as do straw, wood chips and twigs. At the end of the process, screen these out so that they can be reused. In a large pile, lay perforated pipe at the bottom that is open at each end. This draws air into the center of the pile. Turning the pile periodically with a fork or shovel also helps keep it well aerated.

Once the materials are in the pile and the microbes start to work, heat is generated. You may even see steam rising from the pile, and if you open it up, you will get a blast of warm vapors. If this process goes on long enough, it kills weed seeds and pathogens, but few gardeners have compost heaps large enough to retain this much heat. After a week or two, the temperature drops off. This is when you should turn the pile, mixing the materials together and moving material from the top and the outside into the middle of the pile. There is a second heating phase, although not as warm or as long as the first one, after which the pile should be turned again. During the active phase of composting, easily rotted materials are broken down and much of the nitrogen is bound up in stable organic compounds. There is still some ammonium nitrogen along with some volatile organic compounds, and the salt content of the compost may be quite high, so this material may not be completely safe to add to plants at high rates.

Fresh, Rotted or Composted Manure?

We tend to treat these materials as if they are the same, but there are some significant differences.

	Fresh Manure	Rotted Manure	Composted Manure
Nitrogen	Mix of organic nitrogen (slowly available) and ammonium (immediately available)	Almost all organic nitrogen (slowly available)	Mostly organic nitrogen (slowly available), with some nitrate nitrogen (immediately available)
Other nutrients	High in phosphorus and potassium	Fairly high in phosphorus but has lost most of the potassium	High in phosphorus, may have lost some of the potassium during maturation
Salts	Can be high	Low	Can be high, depending on what went into mix
Weed seeds	Most are still viable; raw manure can cause severe weed infestations	Tender weed seeds have died, but more resistant ones are still viable	Proper composting kills weed seeds, but most in home compost are still viable
Benefits to soil structure	Very beneficial	Moderately beneficial	Moderately beneficial
Ease of handling	Chunky; hard to mix evenly with soil	Friable, easily mixed	Friable, easily mixed
Odors	Smelly	Odorless	Odorless

The final phase, which can last for six months or more, is the maturation of the compost. During maturation, the remaining ammonium is converted to nitrate, the volatile compounds either dissipate or are incorporated into stable compounds and the accumulated salts leach away. At this stage, the compost is colonized by specific microorganisms that are beneficial to the soil.

Many companies market inoculants developed to make the composting process faster and more reliable. The question is whether these inoculants make your compost better—or simply more expensive. The first concern is whether the inoculant adds any microbial cells to your compost. You can be fairly certain that when the inoculant left the factory, it contained the reported number of active cells. But leave the inoculant on the car dash for a few hours on a sunny day, and you can be just as certain that none of them will make it home alive.

Even if you follow the directions and add lots of microbes to the compost, this addition still doesn't help if you don't have the right mix of air, water and organic materials. The best compost inoculant can't make up for a poor environment for the microbes.

A compost inoculant *can* make a significant difference when you have done everything correctly, but there aren't enough microbes in the materials going into the compost pile, a situation that doesn't occur very often. If you are worried about a shortage of microbes, the best inoculant you can use is a couple shovelfuls of compost from a neighbor's pile.

NEVER put the following in a compost pile:

- Meat, fat or bones. They attract vermin and smell bad.
- Cat or dog feces. They can transmit diseases or parasites to humans.
- Weeds that have gone to seed. Most home compost piles don't maintain a hot enough temperature over an adequate period of time to kill weed seeds, so when you spread the compost, you spread weeds too. The same goes for perennial weeds that spread vegetatively, such as quack grass or bindweed.

Compost Troubleshooting

As with most activities in the garden, composting doesn't always work the way it should.

Unpleasant Odors

If the compost pile stinks, one of two things may be out of balance: There is too much water, so the pile is anaerobic, or there is too much nitrogen. A wet compost pile has a rancid, sour smell. Turn the pile to create more channels for air to get in, and mix in dry, bulky materials to dry up some of the moisture. Too much nitrogen creates a sewage or ammonia smell. Adding more brown materials helps balance the amount of carbon to the available nitrogen and the microbes in the pile. This takes care of the smelly excess.

Inactive Compost Heap

If the compost is not heating up, there is probably a shortage of nitrogen. Add some green materials so that more microbes can grow to break down the high carbon materials. You may also have to add some water to keep the microbes happy.

Excessive Mold Growth

Mold can occur when the compost gets too dry and is not fully aerobic. It is often a sign that too much of a single material was added at once, rather than being layered in with other organic materials.

Charred Compost

Spontaneous combustion is not common because most backyard compost piles are not large enough to generate the heat needed to start to burn. If you do have a large pile, however, do not let it dry out. The heat created by microbial action drives a lot of moisture out of the pile, leaving dry materials that could catch fire. Watch the compost pile during the active phase, and make sure it stays moist. Turn the pile, and spray the warmest materials with the hose to cool things down.

Is Fast Composting Better Than Slow Rotting?

Each process has its advantages and drawbacks. Waste materials can be left to rot down on their own. The end product is dark-colored, friable and odorless. The slow decay takes up more space in the garden, which may create challenges on some properties. A properly operating composter, on the other hand, breaks down materials as they are added, so it doesn't require a lot of space. The slow-decay process does not kill many weed seeds, so the ones that were viable going into the pile are still viable when they come out. While a large-scale composting operation gets hot enough to kill weed seeds, most backyard composters do not, so both methods fall short in this regard.

In the end, it comes down to personal preference and the amount of space you have available. Unless you are planning to use the material to brew compost teas for disease suppression, the variations in the final product will not make any noticeable difference in your garden soil.

Managing Soils to Minimize Pest and Disease Problems

Pest or disease infestations in your garden depend on three factors:

1. Host: the plant that is susceptible to that particular disease or pest.
2. Inoculum: the pest or disease population at high enough levels to cause harm.
3. Environment: the conditions in which the pest or disease can flourish.

If you eliminate any one of these factors, the infestation will never occur.

When the plant you want to grow is susceptible to a specific pest, your choices are to give up on growing that particular plant, find a cultivar that is resistant to the pest or correct the other two factors.

Crop (or plant) rotation reduces the buildup of diseases or pests by removing the food source, which is the host plant. Choose replacement plants that don't share the same pests. Some insects and diseases are very plant-specific, while others have a broad range of potential hosts.

Another way to reduce disease or insect pressure is to encourage an active and varied soil biology. Beneficial nematodes, fungi and bacteria occupy many of the same spaces in the soil as their pathogenic cousins. If there are a lot of "good bugs," they suppress the population of pests. Healthy soil biology results in a balanced population of grazers and predators that can keep the excessive growth of any organism under control. While this does not eliminate pest pressure, it usually reduces it. Ensure that there is an adequate supply of active organic matter and that environmental conditions favor the growth of "good bugs."

Eliminating disease inoculum by removing the unharvested plant parts disturbs the active soil organic-matter fraction. If you take away organic materials, be sure to replace them by planting a cover crop or adding an organic amendment.

The biggest immediate impact on plant health can be achieved by modifying the soil environment so that it does not favor pests. This is especially true for many plant diseases, which are often caused by fungi. Pathogenic fungi prefer moist conditions, so preventing excessive wetness reduces the virulence of these organisms. When it rains, excess water should be able to drain away either into the soil or across the surface. Eliminate areas where water can accumulate for long periods. Maintaining a strong granular structure is important for good internal drainage on fine-textured soils. Avoid overworking the soil or compacting it.

Whether you use pesticides or manage your garden organically, these methods of reducing disease and insect severity are equally important and valid. It's not unlike how we manage our own health: Although we have antibiotics to treat infections, it is far more effective to avoid the infection in the first place than to treat it afterward.

For More Information

- The USDA-NRCS website has an excellent introduction to soil biology at **soils.usda.gov/sqi/concepts/soil_biology/ biology.html**.
- *Building Soils for Better Crops* (3rd edition) by Fred Magdoff and Harold van Es. Sustainable Agriculture Research & Education, 2010. More suited to farms than to gardens, this text provides an excellent introduction to practical soil science.

Feeding Your Plants

When you would enrich your worne out plantations, if the ground be poor and dry, add well rotted manure prepared and mixt with soil.

— John Reid, *The Scots Gard'ner*

12

A GREAT DEAL OF soil management focuses on the judicious addition of nutrients, so why haven't we addressed this part of the equation earlier? The simple answer is that soil fertility is less likely to limit plant growth than the other soil parameters and is a relatively easy way to effect change. Adding fertilizer when a particular nutrient is lacking produces an immediate response. Correcting poor soil structure or increasing soil life, on the other hand, takes a lot more time and effort.

Determining the right amount of nutrients to add is a matter of understanding what your plants need, what is already in the soil and how to make up the difference. A general recommendation for a single rate of a particular fertilizer or an organic amendment across a wide range of crops and soil conditions is certain to be wrong. At the same time, striving to tailor nutrient applications to the variety of plants in your garden while accounting for the soil conditions in your area would make managing nutrients a major headache. Let's try to find the middle ground, wherein you can be confident your plants are well fed while avoiding the perils of overfertilizing.

Organic Nutrients Versus Mineral Fertilizers

It's a common misconception that organic and mineral nutrients are fundamentally different. In fact, the forms absorbed by plants are identical, but there can be huge variations in how those nutrients are held in the soil and when they become available to plants. Throughout this chapter, options for both organic and mineral sources of various nutrients are provided, along with guidance on how to realize the greatest benefit from each.

How Nutrients Enter Plants

In the 1730s—the early days of scientific agriculture—agricultural pioneer Jethro Tull theorized that plant roots took up nutrients through tiny mouths. By finely pulverizing the soil, he suggested, the farmer could ensure that more nutrients would be available to be "eaten." We now know this theory to be mistaken. Aggressive tillage simply exposes old organic matter within the soil to decomposition, thus releasing nutrients in forms available to plants.

There are two key principles of nutrient uptake: First, most nutrients are absorbed by the roots; and second, the roots absorb nutrients in the form of mineral ions, whether they start out in organic or mineral form. In the previous chapter, we addressed organic-matter decomposition and its role in nutrient cycling. Now let's focus on the movement of the nutrient elements from the soil into the roots.

Root Structure

A root grows when cells divide at the root tip, pushing a cap of cells and mucilage into the tiny spaces between soil particles. This part of the root doesn't absorb nutrients or water—its role is to open a pathway through the soil for the root parts to follow. A few millimeters back from the root tip, specialized cells in the root wall extend fine hairs that absorb nutrients and water from the soil.

A plant must continually grow new roots, since root hairs last for only a few days before dying off. The older roots thicken and differentiate into several different structures, including a tough outer layer and a series of internal pipes called the xylem and phloem. While these older roots anchor the plant and carry water and nutrients to the aboveground part of the plant, they no longer absorb nutrients from the soil.

As roots grow, they release a constant stream of sugars and proteins called exudates into the surrounding soil. About one-third or more of the carbon fixed in the aboveground part of the plant by photosynthesis is exuded from the roots. The root exudates are food for a diverse microbial community that dissolves nutrients from the soil. In the case of mycorrhizae, the return on the investment of sugars is a network of hyphae that effectively extend the reach of the root system, carrying water and nutrients to the plants.

How Nutrients Move to the Root

Direct interception occurs when root hairs come into contact with a soil particle and absorb a nutrient ion directly from that particle. This method accounts for a very small part of the total nutrient uptake by the plant.

Diffusion is a process wherein nutrients dissolved in the soil solution move from an area of high concentration to an area of low concentration. As the root absorbs nutrients from the soil solution, a zone of low concentration is created. Nutrients diffuse into this zone to replace the ones that have been removed, providing a small but continuous supply of nutrients to the root. Nutrients that are held tightly to the soil, like phosphorus and potassium, move mostly through diffusion as they are released from the soil particles.

Mass flow occurs when nutrients in the soil solution catch a free ride on the water molecules absorbed by the root. This is the dominant way that mobile nutrients like nitrate reach the plant.

These processes operate over different physical distances and vary depending on the nutrient. Diffusion does not require direct contact with a soil particle, but it must be within a millimeter or two for there to be enough replenishment of the soil solution to meet the demand of the plant. Mass flow can pull nutrients from a larger zone around the root. Therefore, the placement of nutrients moving by diffusion is more important than for nutrients moving by mass flow. If you put a source of nitrogen on the surface of the soil, you can expect that water from the sprinkler will move it down within range of the roots. But if phosphorus is applied the same way, only a tiny amount will get close enough to the roots to be absorbed, so it is more effective to mix phosphorus into the soil.

The amount of water in the soil influences nutrient movement to the roots. Too much water, and the lack of air means new roots won't grow, so there are few root hairs to absorb nutrients from the soil solution. Too little, and both water and nutrients must follow a more convoluted path around the soil particles to reach the roots, so nutrient uptake slows down. Nutrient uptake by diffusion slows more quickly than by mass flow, so deficiencies of the nutrients that bind tightly to soil particles show up first under dry conditions.

Reading Roots

Since most nutrients are taken up through the roots, consider the roots first if the aboveground parts of the plant do not look healthy. Dig up some plants and examine their roots, or look at the roots when you are transplanting or dividing your perennials. Root systems vary in appearance among plant species, so it's easier to identify abnormal root systems if you have experience looking at healthy roots. We can, however, make some general statements about root systems.

Size
The horizontal spread of a healthy plant should be about the same below ground as it is aboveground. There are exceptions at either end of the moisture spectrum. Some desert-adapted plants have very wide but shallow root systems to capture the infrequent rains, while plants growing in very wet conditions can have similar root systems because downward growth is prohibited by saturated soil.

The depth of the root system is probably not as great as the height of the plant, but expect tall plants to have deeper roots than short ones. Most of the roots occur in the top few inches of soil, but some roots should extend deeper as well. The decline in root number with depth should be gradual rather than sudden. If your garden is in a normal moisture range and the roots appear to be limited in depth, look for possible causes, such as soil compaction or poor drainage.

Other possible causes for a small root system are soil acidity, salt injury and feeding by insects or nematodes.

Shape

Normal roots are generally round in cross section and extend out and down at an angle from the base of the stem. Roots that appear flattened or misshapen or suddenly change direction are a sign of compaction or poor soil structure.

The tips of the roots should taper to a blunt end. If they appear clubbed or twisted, it is usually a sign of trouble, especially if there is discoloration.

Branching

Healthy roots branch out in a fanlike pattern. The degree of branching varies among species, but if there are no side roots or if the root is densely branched like a little Christmas tree, it is considered abnormal. A lack of branching indicates that either the root is traveling through such a tight soil path the side roots have no room to form or something is nipping off the side roots. Excessive branching is more common and indicates that something is stunting the main root or killing it back, and the plant is compensating by putting out more lateral roots. If these are nipped back, the process repeats itself on the laterals, which creates a very dense mass of short, stubby roots. The cause may be extremely low fertility, extreme acidity or a high population of nematodes.

Color

Mature roots may take on dark colors, but young feeding roots should be white to creamy in color. An absence of light-colored roots may indicate that root growth has stalled for some reason, and the plant is not able to absorb water or nutrients from the soil until root growth starts again. This could be a response to weather extremes (too hot, cold, wet or dry). Expect root growth to pick up when the weather moderates.

When combined with the shape and firmness of the roots, discoloration of the feeding roots can be diagnostic of a number of problems in the soil. Many of the symptoms require a hand lens to see clearly.

Color of Feeding Root	Shape	Firmness	Possible Cause
Black	Stunted	Firm	Fertilizer burn
Pink, red or dark	Normal	Mushy	Pathogenic fungi
White to tan; may have dark lesions	Stunted, branched	Firm	Pathogenic nematodes
White to brown	Stunted, branched	Firm	Soil acidity
White to tan	Stunted, few branches	Firm	Insect feeding

Foliar Feeding

Although most nutrients are taken up through the roots, leaves also absorb nutrients. Foliar feeding is a method of spraying liquid fertilizer directly onto the leaves, but there are limits to how much of the plant requirements can be met using this technique. Too high a concentration of nutrient salts in the foliar spray burns the foliage. Diluting the spray and applying more volume results in runoff, preventing absorption. While a foliar application is impractical to supply needed nutrients in large amounts, it can be very effective in overcoming micronutrient deficiencies. A foliar nutrient application also helps a plant look healthier, even though it has little impact on the plant's growth.

Note that foliar feeding is not very effective on plants with thick, waxy leaves or hairy leaves. Waxy leaves block the solution from entering the leaf, while hairy leaves may prevent the solution from reaching the surface of the leaf.

When providing nutrients as a foliar spray, keep the following guidelines in mind:
• Make sure the nutrients are completely soluble and are

intended for foliar application.

- Follow the mixing instructions exactly. The total concentration of nutrients in the spray solution should be less than 1 percent.
- Use some form of surfactant or spreader-sticker with the spray so that it clings to the surface of the leaf and spreads out rather than beading.
- Do not apply foliar spray in bright sunlight, particularly if the weather is hot and dry. The leaf can absorb nutrients only in solution, so as soon as the spray dries on the leaf, it has lost its effectiveness. Evening is ideal, but early in the morning is also satisfactory.
- Apply the foliar spray evenly, just to the point where the spray runs off the leaf.

Nutrients Your Plants Need

There are 17 essential elements for plant growth and reproduction. If any one of these is missing, a plant either does not grow or does not produce flowers or seed.

The first three of these elements—carbon, hydrogen and oxygen—do not, strictly speaking, come from the soil. They are derived from carbon dioxide and water. These are converted by photosynthesis into sugar, which is then incorporated into other compounds as the plant grows.

The remaining 14 elements are divided into three categories:

Macronutrients are needed in relatively large amounts.

Secondary nutrients were included as part of most early fertilizers but are present at much lower concentrations in modern fertilizers.

Micronutrients are needed in tiny quantities.

In addition, there are beneficial elements, which are not essential for plant growth but help plants to grow better or avoid disease.

By eating plants, we get a number of minerals that are essential for our growth and health, including cobalt, iodine, chromium and selenium. These are not required by the plants themselves.

Macronutrients

Nitrogen

Many of the organic amendments we apply to our gardens contain nitrogen, which provides two challenges for the gardener. The first is that the nitrogen may not be available to the plants when they need it. This is particularly true for early short-season crops like lettuce. The decomposition process proceeds very slowly in cold soil, so there may not be enough nitrogen released to meet the needs of the early crops. However, there is more than enough once the soil warms up. The second challenge is that excess nitrogen may be released into the soil, which can lead to vigorous vegetative growth but a shortage of flowers and fruit. If the nitrogen release is later in the season, it can cause nitrate to leach into groundwater during the fall and winter, when there is excess water. If you have a shallow well and a large garden, this could contaminate your drinking water.

Nitrogen is a part of protein. Since chlorophyll, which gives plants their green color, is a protein, more nitrogen usually produces a darker green color. Nitrogen is also associated with abundant vegetative growth. It is mobile within the plant, moving to parts that have the greatest need for it (generally the actively growing tissue, like young leaves).

Nitrogen is needed by plants in relatively large quantities, and inadequate supply results in plants that are stunted and pale green or yellow. Deficiency symptoms show up in the older leaves first. In sweet corn, the lower leaves turn yellow and then brown, from the tip extending down the mid-rib of the leaf in a distinctive "V."

Deficiency/Excess Symptoms

Plants that are short of nitrogen are stunted and pale green to yellow in color. The symptoms first appear in the older tissue near the base of the plant and gradually move upward as nitrogen is translocated to the young, actively growing tissue.

Deficiency symptoms in corn are quite diagnostic, with chlorosis (yellowing) beginning at the tip of the lower leaves and extending along the midrib. In more severe deficiencies, the leaf tissue begins to die and necrosis (browning) follows the same pattern.

Excess nitrogen usually manifests itself as vigorous dark green plants that don't produce flowers or fruit. This is more common in houseplants but is occasionally seen in the garden. Too much nitrogen can also produce plants with tall, weak stalks that break easily or plants with poor resistance to disease.

Excess nitrogen late in the season can be particularly damaging to trees and vines, since it can interfere with the hardening-off process that is essential for winter hardiness.

How Nitrogen Is Held in Soil

Most of the nitrogen in soil is held in organic forms that are not available to plants. As these break down, ammonium is released, which is adsorbed to soil colloids and held there. Under warm, moist conditions, the ammonium is quickly converted to nitrate. Because of its negative charge, nitrate is not held by the soil colloids and goes where the soil water flows. It is carried easily to plant roots by mass flow and also leaches down below the root zone with excess water.

Nutrient Sources

Legumes fix their own nitrogen out of the air. Some of this is available to crops planted the year following a legume.

Compost, especially from livestock manure, provides nitrogen as it decomposes.

Raw manure, particularly broiler-chicken manure, provides large amounts of nitrogen soon after application (raw manure is not permitted in organic production systems).

Urea, ammonium nitrate, ammonium sulfate, calcium nitrate potassium nitrate: All these fertilizer sources of nitrogen are available as soon as they are incorporated into the soil. Urea left on the surface of a moist soil can release ammonia into the air, so it should be either worked into the soil or watered in.

Phosphorus

Lack of phosphorus is a challenge in many organic farming and gardening systems, as the materials that are approved as nutrient sources are expensive and of low concentration. While this is generally not a concern in small gardens, you will still need an external source, whether it is organic or mineral, to replace phosphorus that is removed from your garden.

Among mineral fertilizers, some liquid phosphorus products claim they are more available to plants because they are 100 percent soluble. That may hold true in hydroponic systems, but

This sweet potato plant shows the symptoms most closely associated with phosphorus deficiency: purpling of the leaves, particularly in the lower parts of the plant. Be careful about using this as an indicator of phosphorous deficiency, since there are many other reasons for the leaves to turn purple. Follow up with a soil test to confirm that fertility is to blame before adding more fertilizer.

it is the reaction with the soil rather than the solubility of the fertilizer that determines how much phosphorus reaches the plant. Solid forms of phosphorus are as effective at meeting the needs of the plant as are liquid ones and may be easier to place below the surface of the soil, where the roots can reach it.

The primary role of phosphorus in plants is energy transfer, so it is found in high concentrations wherever there is active growth. It is important for root growth and forms the backbone of DNA, which carries the plant's genetic code.

Deficiency/Excess Symptoms

The classic symptom for a phosphorus deficiency is red or purple leaves. This is also the symptom for almost any stress on the plant so is not terribly diagnostic. Phosphorus deficiency is nearly impossible to diagnose from visual symptoms, since affected plants may only be stunted and darker green than plants with adequate phosphorus.

The only time excess phosphorus becomes an issue is when it induces a deficiency of zinc. Too much phosphorus should be avoided because it is wasteful and can lead to contamination of surface water.

How Phosphorus Is Held in Soil

Plants absorb phosphorus as the phosphate ion, which is extremely reactive in soil. It binds to soil minerals or combines with cations in the soil solution to form insoluble compounds. In alkaline soils, it combines with calcium and magnesium, while in acidic soils, it combines with iron and aluminum. Soil pH has a big impact on phosphate availability, with the greatest availability just below a neutral pH (6.5–7.0). Increased phosphorus availability is one of the benefits of liming acidic soils, but applying too much lime will push the pH up beyond the ideal range and reduce the phosphorus availability.

The reactivity of phosphate in soil has two important consequences for gardeners. The first is that phosphorus mixed into the soil is more available to plants than when it is scattered on the surface. The second is that cool, wet weather, which slows root growth, severely impedes the ability of plants to find and absorb phosphorus from the soil. Placing the nutrient in a concentrated band near the seedlings helps overcome these early phosphorus deficiencies.

Nutrient Sources
Bonemeal, composted manure, raw manure, particularly swine and poultry, triple superphosphate (monocalcium phosphate), monoammonium phosphate or diammonium phosphate, ammonium polyphosphate (liquid).

Potassium

When early settlers burned and cleared the forests of North America, they mixed the remnant wood ashes with water and extracted the "pot-ash," which was used as a fertilizer. Although wood ashes are a source of potassium, you must take some precautions when applying them to your garden. At high rates, the ashes are caustic, so don't apply them immediately before planting or to growing plants. Never use ashes from pressure-treated wood, since the toxic compounds used to preserve the wood are concentrated in the ashes.

The main role of potassium is to regulate the plant's water balance. Cells that are short of potassium are unable to maintain their turgor (firmness), which can lead to poor drought tolerance and poor disease resistance. Potassium does not get bound up in compounds in the plant but stays in the cell solution as the potassium ion.

Deficiency/Excess Symptoms
Potassium is mobile in the plant, so deficiency symptoms show up first in the lower, older leaves as chlorosis (yellowing), followed by necrosis (browning), starting at the margins of the leaves. Eventually, the leaf edges become ragged as dead tissue breaks away.

High levels of potassium are not, on their own, harmful to the plant. But if soil magnesium levels are low, high potassium levels can compete for plant uptake and induce a magnesium deficiency. Excessive applications of potassium fertilizers (or manure containing high levels of potassium) can lead to high

Like nitrogen, potassium deficiency tends to show up in the older leaves. The leaf margins turn yellow and then brown. In severe cases, the leaf margins become ragged as the dead tissue breaks away. Potassium deficiency shows up more frequently with dry weather and compacted soils.

levels of salts in the soil solution. This causes fertilizer burn if the weather is dry.

How Potassium Is Held in Soil

The potassium ions are held to the clay and organic matter by cation exchange. This potassium is readily available to plants. Some is also held between the layers of clay minerals. The structure of the clay layers is a bit like an egg carton, and the potassium ions fit in the spaces within this matrix. When the clay layers close around the potassium ions, the electrical charges are close enough to hold the layers together, trapping the potassium ions inside.

The plant root absorbs potassium from the soil solution by swapping the potassium ions for hydrogen ions. This creates a zone of low potassium concentration next to the root, and more potassium diffuses into this zone while the hydrogen diffuses out. Dry weather can, therefore, limit potassium movement to the plant. The water films around the soil particles follow a more tortuous path, so the potassium ions must travel a greater distance.

Nutrient Sources

Compost (particularly from livestock manure), wood ashes, raw manure, especially from beef or dairy cattle, muriate of potash (potassium chloride), potassium sulfate, potassium magnesium sulfate.

Secondary Nutrients

Calcium

Calcium disorders, like blossom end rot, occur when a dry period is followed by an abundance of water. When the soil becomes dry enough that moisture uptake is restricted, the water carrying dissolved calcium ends up in the leaves, because that is where most of the water is transpired. If there isn't much transpiration taking place, the developing fruit is shortchanged on calcium and the blossom end of the fruit suffers the most.

The cells developing in this region have low levels of calcium in the cell walls and so lack strength. When the moisture supply increases, from a rainfall or irrigation, the fruit expands rapidly as the cells absorb water. The weakened cells at the blossom end burst open during this expansion, exposing the fruit to decay organisms.

Theoretically, a foliar application of calcium could prevent blossom end rot, but to be effective, it must be applied directly to the developing fruit. Since the fruit at this stage is small and hidden beneath the leaves, this is completely impractical. Similarly, soil-applied fertilizers used to improve the calcium supply to the plant are unlikely to work, since the problem is movement within the plant rather than supply from the soil. The best way to prevent blossom end rot is to water the plants consistently. Unless there is an extreme drought, a well-structured soil with high organic matter can accomplish this. Simply water plants deeply a couple of times each week.

Calcium fertilizers can prevent calcium disorders in one vegetable: the potato. Because the tuber develops underground,

Calcium deficiency always shows up at the growing point of the plant and is more often caused by poor calcium transport within the plant (because of moisture stress) than because of a shortage in the soil. The most common example of this is blossom end rot in tomatoes, but it can also show up as tip burn in leafy vegetables. In cabbage, the growing point is hidden within the head, so it is not apparent until after harvest.

it can absorb calcium directly from the surrounding soil if the levels are high enough. Adding some gypsum or calcium nitrate to the soil at hilling time improves tuber quality in low-calcium soil.

Calcium is a component in some enzymes, but most of the calcium is tied up in the cell walls, where it helps give the cells rigidity and strength.

Deficiency/Excess Symptoms

A shortage of calcium in the soil is rare, since every soil that has adequate pH for plant growth has adequate calcium too. However, there are disorders in specific parts of plants that are caused by calcium deficiency. These are the result of an irregular moisture supply. The most common example of this is blossom end rot in tomatoes or peppers, as well as tip burn in lettuce, bitter pit in apples and black heart in celery and potatoes. Calcium moves only from the roots to the top of the plant (it is immobile in the rest of the plant), so any symptoms of deficiency show up in the newest tissue at the top. Foliar sprays of calcium are ineffective because calcium cannot move from one part of the plant to another.

While there is no detrimental effect from a high level of calcium in the soil, it is often associated with a high pH level, which can reduce the solubility of many micronutrients. If you live in an area with alkaline parent material, a high level of calcium can be an indicator that you are growing plants in subsoil and may be faced with poor soil structure from a lack of organic matter.

How Calcium Is Held in Soil

Calcium ions are held on the soil colloids by cation exchange, but there is also a significant amount of calcium in the soil solution. Most of the calcium that plants absorb is by mass flow in the water taken up by the roots. Some plants, like pine trees, exclude part of the calcium from the water they absorb. This calcium can build up over time as a cylindrical deposit of calcium carbonate (limestone particles) around the roots.

In alkaline soils, a significant quantity of calcium may be

present in the form of calcium carbonate. As acidity is added to the soil, this gradually dissolves, releasing calcium into the soil solution.

Nutrient Sources

Manure from egg-laying hens is high in calcium, because their feed is supplemented with calcium to produce hard eggshells. Agricultural lime, while not strictly an organic compound, is allowed in all organic systems; it's a good source of calcium and helps raise soil pH.

Agricultural lime (calcium carbonate), gypsum (calcium sulfate), calcium nitrate, calcium chloride.

Magnesium

Excessive applications of potassium can induce a magnesium deficiency in plants, even when there is an adequate supply of magnesium in the soil. Overenthusiastic applications of fertilizer, manure or compost can sometimes lead to interactions that hinder, rather than help, plant growth.

Magnesium is the heart of the chlorophyll molecule, so it is critical to the process of photosynthesis. It also plays a role in protein synthesis and in activating some enzymes.

Deficiency/Excess Symptoms

Magnesium is mobile within the plant, so any deficiency symptoms are usually seen on the older plant tissue first. Symptoms vary among plant species but often show up as chlorosis (yellowing) between the veins of the leaf, which may be accompanied by a reddish or purplish color. In some species, brown spots develop in the middle of these interveinal areas. Magnesium toxicity has not been identified as a problem.

How Magnesium Is Held in Soil

Magnesium ions can be found in the soil solution and adsorbed on soil colloids, as well as in particles of dolomitic lime in the soil. Coarse-textured acidic soils are often deficient in magnesium.

Magnesium deficiency appears as mottling or striping between the veins of the leaves, as seen here on a corn leaf. This shows up most often in sandy or acidic soils.

Nutrient Sources

Dolomitic limestone (calcium magnesium carbonate), while not strictly an organic compound, is allowed in all organic systems.

Dolomitic limestone, epsom salts (magnesium sulfate), potassium magnesium sulfate.

Sulfur

Most of eastern North America has not needed sulfur fertilizers for the past century because enough sulfur has been added to the soil by acid rain. The highest levels of deposition are around the Great Lakes and in the Ohio Valley. But since the implementation of stringent emission standards on power plants and industry, the levels have been gradually declining, so sulfur additions will likely be needed in future. The western half of the continent is upwind of most of the industrial sulfur emissions and so has benefited from sulfur all along.

Sulfur is part of two of the 21 amino acids that form protein. It helps develop enzymes and vitamins, aids in seed production and is needed for chlorophyll formation. Sulfur also adds color, flavor and a distinctive odor to plants such as garlic, onions and cabbage, and it puts the heat in horseradish.

Adding sulfur to plants can make them darker green in color, even if they have adequate sulfur for growth. Some fertilizers include sulfur specifically for this cosmetic effect.

Deficiency/Excess Symptoms

Sulfur deficiency looks similar to nitrogen deficiency, except that the symptoms show up on the entire plant rather than showing on the lower leaves first, because sulfur is not mobile within the plant. Plants are stunted and pale green and may have chlorosis (yellowing) between the veins. Members of the brassica family may also show cupping and purpling of the leaves. Sulfur toxicity has not been identified as a problem.

How Sulfur Is Held in Soil

The largest reserve of sulfur is in the soil organic matter. As this breaks down, sulfate ions are released into the soil solution and taken up by the plants. Because it is negatively charged, sulfate

Sulfur deficiency can be easily confused with deficiencies of nitrogen (yellowing and stunting of the plant) or magnesium (yellowing between the veins). The difference with sulfur is that it is usually the entire plant that is affected, as seen in the lettuce in the pot on the right.

is not held by soil colloids and can leach out of the soil when there is excess moisture. In a dry environment, the sulfate may combine with calcium to form gypsum.

Sulfur deficiency often occurs in coarse-textured soils that are low in organic matter. Soil tests are not a reliable indicator of whether there is adequate sulfur in the soil. Organic-matter content is likely a better indicator of sulfur supply.

Elemental sulfur is not usable by plants until it has been converted to sulfate by specialized bacteria that oxidize sulfur, gaining energy in the process. This conversion takes time, so elemental sulfur that is applied to the soil may not be available to plants until the following year.

Nutrient Sources

Most manures or composts contain sulfur, but it may not be at a high enough concentration to overcome a sulfur deficiency. The nitrogen in these materials can compete with sulfur for uptake by the plants.

Elemental sulfur (for next year's crop), ammonium sulfate, potassium sulfate, gypsum (calcium sulfate), potassium magnesium sulfate.

Boron deficiency is not very common, but where it does appear, it is generally in the flowers or the fruit. Members of the cabbage family have the highest requirements for boron, and the poor curd formation in this cauliflower is a typical symptom.

Micronutrients

Boron

Boron is important for the strength of cell walls, fruit set, seed development and carbohydrate and protein metabolism.

Deficiency/Excess Symptoms

Plant requirements and tolerance for boron vary widely. It is not mobile in the plant, so deficiency symptoms are generally seen in the newest tissue.

Root crops may exhibit cracking of the root or hollow root centers (e.g., watery core of rutabaga). Cole crops may show hollow stems and deformed buds.

Most plants are very sensitive to excess boron, so boron toxicity is a more common problem than boron deficiency. Symptoms of boron toxicity include bleaching of plant tissue and browning along leaf margins.

How Boron Is Held in Soil

Much of the boron reserve is in the soil organic matter, which releases borate when it breaks down. Since borate does not carry any electrical charge, it readily leaches out of the soil. The amount of borate in the soil is controlled by the breakdown of organic matter and the rate of leaching. Sandy soils, low in organic matter, are most likely to show boron deficiency, but only when dry conditions slow the breakdown of organic matter.

Nutrient Sources

Compost and manure both contain a variety of micronutrients. Borax (sodium tetraborate).

Chlorine

Some farmers and gardeners believe that applying anything with chloride will sterilize the soil. This mistaken belief is understandable, since chlorine is associated with some pretty nasty compounds. Chlorine gas kills bacteria, which is why it is injected into drinking-water systems. Organochlorine compounds are among the most toxic and persistent insecticides. These are both very different, however, from the chloride ion that is essential for all life. To put this in perspective, water-treatment plants typically aim for a concentration of 1.7 parts per million of chlorine (the element) to disinfect drinking water, while seawater, which teems with all manner of life, has a chloride (the ion) concentration on the order of 20,000 parts per million.

Present in the plant as the chloride ion, chlorine plays a role in water balance within the plant.

Deficiency/Excess Symptoms

Chloride deficiency is rare enough that specific deficiency symptoms have not been identified for any crop except wheat, where speckling on the leaves may be seen.

With the exception of some tree fruits and cane fruits, most plants are tolerant of high levels of chloride. Since high chloride can be associated with high salt concentrations in the soil, some injury may be caused by the salts rather than the chloride.

How Chlorine Is Held in Soil

Chloride occurs either in the soil solution or within a living organism. It does not form any compound that isn't extremely soluble in water, so chloride minerals (sodium chloride, potassium chloride) are present only under arid conditions.

Nutrient Sources

Muriate of potash (potassium chloride).

Copper is sometimes bound tightly enough to soils with high organic matter that it is not available to plants. This shows up as leaf tips that become floppy and twisted.

Copper

Copper plays a role in chlorophyll production and in some enzyme reactions.

Deficiency/Excess Symptoms

Copper deficiencies are most common in organic soils and in very sandy alkaline soils. Copper-deficient carrot roots are pale-colored rather than rich orange. Onions may have dieback of the leaf tips, which then curl like a pig tail. Lettuce leaves lose their firmness, and the stems are bleached.

Copper toxicity may show up as a deficiency of iron or zinc, as copper competes for uptake of those micronutrients. It can also appear as browning along the edge of the upper leaves, a general symptom of micronutrient toxicity. The water flowing through the xylem carries the micronutrient as far as it can go. As water evaporates, copper accumulates at the edge of the upper leaves, until the concentration is high enough that injury to the tissue can occur.

How Copper Is Held in Soil

Copper ions are found in the soil solution or adsorbed to clay and organic-matter colloids in the soil. It binds very tightly to organic matter, so although an organic soil may have a large supply of copper, very little of it is available to plants.

Nutrient Sources

Copper sulfate

Iron

Iron is included in many fertilizers for turf and ornamental plants. While it does not improve the growth of the plants, it gives the foliage a darker green appearance.

Iron is a catalyst in the formation of the chlorophyll molecule. It also plays a role in plant respiration and protein formation.

Deficiency/Excess Symptoms

Iron deficiency generally shows up as stunting of the plant and yellowing between the veins of the upper leaves. Iron is

Iron deficiency appears as yellow or white discoloration between the veins, most often on the youngest growth. It can show up on annual broadleaf plants like tomatoes or on tree fruits such as the apple shown here.

immobile within the plant, so symptoms appear on new tissue. Iron deficiency is a common cause of poor growth in blueberries and azaleas that are grown in alkaline soil. Broadleaf plants seem to be more susceptible than plants in the grass family.

Iron toxicity is rare and is restricted to acidic soil with excess moisture. Symptoms often show up as a deficiency of another micronutrient, such as manganese, but there have been reports of stunted plants with leathery, dark green leaves.

How Iron Is Held in Soil

Iron is abundant in most soils, but it is tied up in forms that plants cannot use. Plants take up the ferrous ion, which is found in the soil solution or adsorbed to soil colloids. Much more iron is in the insoluble ferric form and is precipitated out as iron oxide (rust). The reddish brown color of many soil horizons is iron oxide, but this does not contribute very much to the supply of iron for plants. The solubility of iron varies with soil pH and aeration. Iron availability increases rapidly as the soil pH drops and in waterlogged soils.

Nutrient Sources

Ferrous sulfate

Manganese deficiency can look very similar to iron deficiency, with the tissue between the veins becoming pale green, then yellow or white, while the veins stay dark green. It may be subtle or quite striking, as in the bean leaf. Symptoms appear on the new leaves, but the next leaves to emerge may be normal as the plant reaches a new supply of manganese in the soil, leaving the affected leaves in the middle of the plant.

Manganese

Do not confuse manganese with magnesium. Despite their similar names, these two elements have very different roles in the plant, and they are in short supply under almost opposite conditions. If your soil has a manganese deficiency and you add dolomitic lime because you have mistaken the problem for a magnesium deficiency, you will worsen the manganese deficiency by raising the soil pH.

Manganese is involved in photosynthesis and chlorophyll production and in activating some enzymes within the plant.

Deficiency/Excess Symptoms

Manganese deficiency in many broadleaf plants is distinctive. The entire leaf becomes yellow or even white, while the veins remain dark green. Manganese is not mobile within the plant, so a deficiency appears on the upper leaves. In some cases, the supply of manganese from the soil improves, and while the symptoms of the temporary deficiency can still be seen in the middle of the plant, the new foliage is normal. Manganese-deficient beets show russeting, curling and stunting of the foliage.

Manganese toxicity occurs in plants growing in very acid soil (pH below 5). Symptoms are similar to those of copper toxicity (necrosis of leaf margins). Excessive manganese in low-pH soil causes some fruit trees to develop "measles" on the bark (raised pimples underlain by brown spots). Liming the soil usually cures manganese toxicity in future seasons.

How Manganese Is Held in Soil

Most soils contain a lot of manganese, but very little is in the ionic form that is taken up by plants. When the pH is high or the soil is well aerated, manganese forms insoluble compounds. It also tends to be bound tightly to organic matter, so manganese availability can be very low in organic soils. Some is released as organic matter decomposes.

Because of the influence of aeration on manganese solubility, manganese availability can vary through the growing season. There is more air in the soil when it is dry, so dry conditions

Unlike any of the other micronutrients, molybdenum is less available in acidic soils, so adding lime can often prevent molybdenum deficiencies. The growing point is affected, so deficiency symptoms, as shown in these broccoli plants, may show up as deformed leaves ("whiptail") or as a hollow core within the stem.

can lead to a temporary shortage of manganese. Cold weather aggravates this condition, since it slows the decomposition of organic matter. A heavy rain or even compacting the soil to drive out some of the air increases the solubility of manganese.

Avoid overliming the soil—it's a sure way to induce a manganese deficiency. Do not add lime unless you have taken a soil test to confirm that the soil pH is low.

Nutrient Sources

Manganese sulfate. Because a shortage of manganese is often a tie-up in the soil rather than an actual shortage, a foliar application of manganese may be more effective than adding it to the soil.

Molybdenum

Molybdenum plays an important role in nitrogen metabolism within the plant, as well as nitrogen fixation in legumes.

Deficiency/Excess Symptoms

Molybdenum is required in such small quantities that deficiencies in garden plants are rare, although in some areas in the Great Plains, molybdenum seed treatments are necessary on legumes to ensure adequate nitrogen fixation. One group of vegetables that does show molybdenum deficiency is the cabbage family

Zinc deficiency symptoms appear most frequently in corn, although other garden vegetables may suffer yield loss without any noticeable symptoms. The growing point is affected, so new tissue takes on a pale green to white appearance, with some sections of the leaf becoming almost transparent. This bleaching extends across most of the base of the leaf blade, rather then being confined between the veins.

(brassicas), where it appears similar to nitrogen or sulfur deficiency. Cauliflower develops a distinctive upward curling of the upper leaf margins, forming a "whiptail."

Molybdenum toxicity is even rarer than molybdenum deficiency, although it has been known to occur at very high soil concentrations and high soil pH. Stunting of the plants and necrosis of the leaf margins are typical toxicity symptoms.

How Molybdenum Is Held in Soil

Molybdenum is unique among the micronutrients because it is more available under alkaline conditions. At low soil pH, much of the molybdate (the plant-available form) is bound to iron and aluminum. Liming acidic soils releases some of this molybdate into the soil solution, where it can be taken up by plants.

Nutrient Sources

Molybdenum trioxide (applied as a seed treatment), sodium molybdate, ammonium molybdate (applied as a foliar spray).

Nickel

Because of the tiny amounts required, nickel has only recently been identified as an essential element. It is important for nitrogen transformations within the plant.

Deficiency/Excess Symptoms

There are no documented examples of nickel deficiency in field-grown plants. High levels of nickel from industrial contamination can induce deficiencies of zinc or iron.

How Nickel Is Held in Soil

The nickel ion that is absorbed by plants is held on the soil colloids.

Nutrient Sources

None required.

Zinc

Zinc plays a role in carbohydrate and chlorophyll production and in seed production. It is important in early plant growth.

Deficiency/Excess Symptoms

Zinc is relatively immobile in plants, so deficiencies show up first on the youngest leaves. In severe deficiencies, the leaves of corn seedlings emerge from the whorl completely white ("white-bud"). Fruit trees and strawberries often show chlorosis (yellowing) of the young leaves and a "halo" of green around the edges. Onions may be stunted, with twisted, yellow-striped foliage.

Plants are relatively tolerant of high zinc levels, but if toxicity occurs, it appears as a deficiency of one of the other micronutrients.

How Zinc Is Held in Soil

Zinc ions are held on the soil colloids.

Nutrient Sources

Zinc sulfate

Responses of Garden Crops to Micronutrient Fertilizers

	Manganese	Boron	Copper	Zinc	Molybdenum
Asparagus	low	low	low	low	low
Beans	high	low	low	high	medium
Beets	high	high	high	medium	high
Blueberries	low	low	medium		
Broccoli, Cauliflower	medium	high	medium		high
Cabbage	medium	medium	medium	low	medium
Carrots, Parsnips	medium	medium	medium	low	low
Celery	medium	high	medium		low
Cucumbers	high	low	medium		
Lettuce	high	medium	high	medium	high
Onions	high	low	high	high	high
Peas	high	low	low	low	medium
Peppers	medium	low	low		medium
Potatoes	high	low	low	medium	low
Radishes	high	medium	medium	medium	medium
Spinach	high	medium	high	high	high
Sweet Corn	high	medium	medium	high	low
Tomatoes	medium	medium	high	medium	medium

If the micronutrient concentration in the soil is low and micronutrient fertilizer is applied, high-response crops often show improved growth or vigor. Medium-response crops are less likely to respond, and low-response crops usually do not respond, even at the lowest micronutrient levels.

Beneficial Elements

Elements in this group are not essential, because plants can grow and reproduce without them, but they do help plants grow better and resist disease.

Cobalt

Although it has no direct role in plant growth, cobalt is necessary for nitrogen fixation in legumes. It also helps in the formation of vitamin B_{12} in plants, which is extremely important for human and animal nutrition.

Sodium

When potassium is in short supply, sodium can replace some (but not all) of its functions in regulating water within the plant. High levels of sodium can be toxic to plants, although it is unclear whether this is due to direct toxicity by sodium or the osmotic effect of high salt concentrations. Tolerance to high sodium concentrations varies widely, so check with your local agricultural office regarding sodium tolerance of plants adapted to your area. Purdue University, in Indiana, has published an extensive list of trees adapted to the Midwest (www.extension. purdue.edu/extmedia/id/id-412-w.pdf).

The biggest impact of high sodium levels is on soil structure. Sodium ions tend to push clay colloids apart, dispersing them rather than holding them together in a stable structure. The poor soil structure that results can impede water drainage, root growth and even seedling emergence.

Silicon

While silicon makes up a large part of most soils (quartz sand, for example, is almost pure silicon dioxide), only a small part is soluble and available to plants. Within the plant, silicon strengthens cell walls, which, in turn, improves the ability of the plant to grow upright. It can also increase the resistance of the plant to root parasites and leaf and root diseases and protect against some nutrient toxicities. There is no evidence yet for any benefits from silicon fertilizers for garden plants, since even small amounts of soluble silica in soil meet plant requirements. Silicon fertilizers have, however, been shown to be beneficial in hydroponic production.

Soil Testing and Plant Analysis

When it comes to applying fertilizer, gardeners typically fall into one of three categories. The first is to carry on as usual if it seems to be working. The second is to buy something off the shelf at the garden center and apply it according to the general directions. The third is to measure what is actually in the soil or plant

Making Sense of the Labels on Fertilizer Containers

Every fertilizer product features a label with a series of numbers. To the uninitiated, these numbers may seem like a mysterious code. Let's demystify them.

Each number represents the percentage of a particular nutrient in the fertilizer. The first three numbers stand for nitrogen, phosphorus and potassium. The slight complication is that there are specific definitions for each nutrient:

- Nitrogen is expressed as the total amount of elemental nitrogen (N) in the material. Organic materials should also include an estimate of how much nitrogen is available to plants in the year the material is applied as the organic compounds break down and release mineral forms of nitrogen.

- Phosphorus is expressed as the amount of phosphorus pentoxide (P_2O_5) available to plants. It is determined by dissolving the fertilizer in a weak acid, mimicking

what plant roots do. Phosphorus pentoxide is not actually found in the fertilizer, nor is it a form that would be available to plants. Its use goes back to the days of determining the phosphorus content of bonemeal by burning it, then weighing the ash. The weight of phosphorus pentoxide is 2.29 times the weight of elemental phosphorus.

- Potassium is expressed as the amount of potash (potassium oxide, or K_2O) that will dissolve in water. The weight of potash is 1.20 times the weight of elemental potassium.

- The remaining nutrients are expressed as the total amount of each nutrient element in the material, with the chemical symbol for the nutrient following the number. All these numbers will not add up to 100 percent, since they do not account for the elements that make up the chemical compounds in the fertilizer.

and add nutrients to make up for any shortfalls. Depending on the nutrient in question, each approach has its advantages and limitations.

The status quo approach works well in many cases, as long as you are applying some nutrient but not too much. It assumes there is already a reasonable level of fertility in the soil, and you are simply maintaining it. The drawback to this approach is that problems of deficiency or excess can sneak up on you, and you won't know until plant growth begins to suffer.

Off-the-shelf products at garden centers are generally high-quality sources of nutrients. The trouble is that there is no relationship between the directions for applying these materials and the nutrients present in your soil. The recommendations typically define the maximum rate that can be applied without causing injury to plants, rather than the rate that will meet the

Too much nutrient can be just as harmful as too little. Excessive concentrations of nutrients near tender roots can burn them, the same way that holding a lit match to them would.

specific needs of the plants in your soil. Problems can arise when you buy multiple products and apply each of them to the maximum rate on the label.

By measuring the nutrients that are already in your soil, you can target your nutrient applications more precisely, heading off problems before they arise. This can be accomplished by analyzing either the soil or the plant tissue.

Soil Testing

A soil test is most useful for measuring nutrients that are retained in the soil and taken up by plants in large amounts. It can also determine soil characteristics such as pH, salinity and organic-matter content. A soil test can provide numbers for concentrations of nitrogen and sulfur, but these numbers should be treated with caution, since they represent the value on the specific day the soil sample was collected and are likely to vary widely as the weather changes. A soil test for micronutrients is not as accurate as the test for macronutrients—the quantities are so small that the test results may not relate to the levels actually available to plants. Check with your local agricultural office to find out which tests are meaningful for your area.

Collecting the Sample

The soil sample should represent all the soil in your garden. Collect at least 10 cores from different parts of the garden, and mix them together thoroughly. Put approximately 1 pound (0.45 kg) of this sample into the container for shipment to the lab. (You can also use a shovel or spade to take narrow slices of soil from different parts of the garden, but you will end up collecting far more soil to get your sample.) Do not mix cores together from parts of the garden that have had very different treatments in the past. For these, send separate samples.

Choosing the Lab and the Tests

Some states and provinces have programs that assess the performance of soil test labs and can provide lists of recommended labs. If your area does not have such a program, look for a lab that is enrolled in an external quality-assurance program, such as the North American Proficiency Testing Program. A list of participating labs can be found on the program website (www.naptprogram.org).

The choice of soil test extractant for each nutrient is as important as is the choice of lab. This is particularly true for phosphorus, because a number of extractants have been developed for different soil conditions. Each extractant provides results on a different scale (not unlike the Fahrenheit and Celsius scales for temperature), so the results cannot be used interchangeably. Check with your local agricultural office for the tests recommended for your area. As a general rule, the acidic extractants (Bray, Mehlich 3, Kelowna) are used where soils are predominantly acidic or neutral in pH, while the alkaline extractants (Olsen [sodium bicarbonate], ammonium bicarbonate) are used in areas dominated by alkaline soils. A lab that is geographically close to you is more likely to use tests that are appropriate for your area.

Each garden soil test should include soil pH, available phosphorus, potassium, calcium and magnesium, organic matter and electrical conductivity (salt content). Other tests may be appropriate for your area, but in many cases, they simply generate numbers that don't mean very much.

Interpreting the Results

In my experience, soil tests for home gardens tend to be extremes. The results indicate that nutrients are either extremely low (a new garden established on the subsoil left behind where a subdivision has been built) or extremely high (a garden with a long history of fertilizer, manure and compost additions). Within this context, trying to target exactly what a specific plant may need is a futile exercise. Aim, instead, to establish your garden soil somewhere in a middle range, where nutrient deficiencies will not limit the growth of your plants but are not so high that problems with toxicities or adverse interactions will appear.

When you receive your soil test report, look first at which specific extractants (Bray, Mehlich 3, Olsen, etc.) have been used for each nutrient. Then check which units have been used. Some labs report nutrient results in parts per million (ppm), which is the same as milligrams per kilogram (mg/kg), while others report results in pounds per acre (lb/ac). Since an acre of topsoil weighs approximately 2 million pounds (about 2 million kilograms in a hectare), the readings in pounds per acre are double the values in ppm.

All those numbers on a soil test report can be intimidating, and the hardest thing for most gardeners is to sort out the few meaningful numbers without being distracted by the many superfluous numbers. Concentrate on the key pieces of information in the report, and try to ignore the rest. Don't be afraid to ask someone for help if you are confused by the report. The local agricultural office, a master gardener and the experienced staff at a garden center are all good sources of information.

Interpreting Soil Test Values

		Low/Deficient	Optimum	High/Excessive
Soil pH		<5.5	6.5–7.5	>8.0
Electrical Conductivity	ms/cm*	—	<0.25	>0.5
	ds/m†	—	<2	>5
Phosphorus				
Bray	ppm	<20	30–100	>150
	lb/ac	<40	60–200	>300
Mehlich 3	ppm	<30	40–150	>200
	lb/ac	<60	80–300	>400
Olsen (sodium bicarbonate)	ppm	<15	20–60	>100
	lb/ac	<30	40–120	>200
Kelowna	ppm	<20	25–100	>125
	lb/ac	<40	50–200	>250
Potassium				
Ammonium Acetate	ppm	<90	150–300	>750
	lb/ac	<180	300–600	>1,500
Mehlich 3	ppm	<100	150–300	>750
	lb/ac	<200	300–600	>1,500
Calcium				
	ppm	<300	>300	—
	lb/ac	<600	>600	—
Magnesium				
	ppm	<50	>100	—
	lb/ac	<100	>200	—
Organic Matter				
Sandy soil	%	<2	>3	—
Loam soil	%	<4	>6	—
Clay soil	%	<6	>8	—

*ms/cm = millisiemens per centimeter, measured in a 2:1 soil–water paste

†ds/m = decisiemens per meter, measured in a saturated soil extract

These values are rough guidelines only. Appropriate ranges vary for different plant species. The ranges given here are higher than you might see if you looked at a similar chart for field corn or wheat, since many fruit and vegetable plants require higher levels of fertility for optimum growth. Adding nutrients when the soil test is above the optimum range will not likely improve plant growth and may increase the risk of polluting ground or surface water.

Plant Analysis

Plant tissue analysis enables you to gauge the concentration of nutrients that are reaching your plants. It is particularly useful for micronutrients, for which soil tests are not accurate, and mobile nutrients, such as nitrogen and sulfur. Even so, it represents only one piece of information to guide your decisions about nutrient applications. Plant analysis can't tell you whether the absence of a specific nutrient is due to a shortage in the soil itself or to a restricted root system that is preventing the plant from extracting soil nutrients. Nonetheless, it can be an extremely useful diagnostic tool when part of the garden is doing poorly for no obvious reason. Always collect samples from both the good and the poor areas so that you can compare the results.

Collecting the Sample

The concentration of nutrients in the plant constantly changes as the plant grows and matures, as does the ratio between different nutrients, and varies from one part of the plant to another. Tables that provide expected ranges for nutrient concentrations are based on samples of specific plant parts collected at a particular stage of growth. Match your sample collection time to an interpretation table. The most common sampling recommendation is to collect the top leaf when it is fully open, just as the plant is beginning to flower. In a few species, you're required to take different plant parts. With grapes, for instance, you must provide a sample of the petioles (leaf stems) rather than the leaf.

Be sure to keep tissue samples clean. Even a small amount of soil on the tissue can significantly skew the results. If you cannot take the samples to the lab immediately, store them in paper bags rather than plastic so that moisture can escape. If the sample becomes moldy before it is analyzed, the results will be meaningless.

Choosing the Lab and the Tests

To choose a lab for your plant tissue analysis, follow the same guidelines as for your soil tests. Most soil testing labs provide both services. They also offer packages for plant tissue analysis that include the most commonly deficient nutrients.

Interpreting the Results

Unless there is a severe deficiency of a particular nutrient, interpreting plant tissue results can be a complex and frustrating exercise. Many factors can influence the concentration of nutrients in the plant. The lab generally provides a comparison of your tissue results to expected ranges, but a reading that is "below average" should be treated with caution. These ranges are usually based on the average nutrient concentrations in a large number of samples, with no consideration as to whether low concentrations have an adverse effect on plant growth. In many cases, the best use of a tissue test is to confirm that nutrient deficiency is not the cause of poor growth in your plants. You can then focus your efforts on uncovering other factors that may be limiting plant growth.

Correcting Nutrient Deficiencies

Once you have determined that your plants require extra nutrients, how should you provide those nutrients? There are two approaches, which are often treated as if they are mutually exclusive. In fact, they should be complementary. The first is to improve the availability of the nutrients that are already in the soil; the second is to add extra nutrients to the soil in a readily available form.

Making Better Use of What Is Already There

This is the heart of the organic approach to gardening and is extremely effective as long as there is a reasonable supply of nutrients in the soil. Using better soil management to improve the cycling of nutrients from the soil to the plants and back again is not just for organic gardeners. It works equally well if you use fertilizer along with organic sources of nutrients.

Building soil tilth is the first step to help plants extract more nutrients from the soil. In deep, friable soils, plants can develop extensive root systems, which allow them to pull nutrients from a larger volume of soil and to explore the soil within that volume more fully. When the total root length is double or triple that of a plant growing in soil with bad tilth, the total nutrient uptake is significant, even if the concentration of nutrients along each inch of root length is not very high.

The second step in making nutrients more accessible to plants is to ensure that conditions are favorable for a healthy soil biota. An active population of organisms rapidly cycles nutrients from plant residues back into available forms for plant uptake. Soil microbes also hasten the breakdown of minerals in the soil, releasing the nutrients locked within. By providing adequate food and the desirable physical environment that a soil in good tilth creates, it's possible to support the soil's biological community.

However, these processes work only if you cycle the plants in your garden back into the soil. Either leave all the unharvested parts of the plants in place to gradually decay or remove them to a compost heap and return them as finished compost. The key is to close the cycle as much as possible. Otherwise, you are mining the nutrients from your soil, and eventually, they will be exhausted.

Even with the most efficient recycling of nutrients, it is impossible to return everything to the soil. The vegetables we harvest and the flowers we cut for the table remove nutrients that are not returned to the garden. Every nutrient cycle experiences some losses, whether to the air or deep in the soil where the plants do not reach. Ultimately, we need to add some nutrients to the soil to maintain healthy plant growth.

Adding Nutrients—How Much?

The most common mistake gardeners make is to apply more nutrients than the plants can use. Whether it's compost or fertilizer, too much is still too much. The excess either is lost into the environment or lingers to the point where it can become a problem for future crops if the overapplication continues.

The problem is that home gardens are relatively small, and the nutrient sources we use are fairly concentrated. Let's work through the math for a couple of examples. We'll assume that the soil test shows moderate fertility levels, so the recommendation is for 90 pounds per acre (100 kg/ha) each of nitrogen, phosphate and potash. If we use a 10-10-10 fertilizer, this translates to about 1 ounce (28 g) of product (a little more than a tablespoon) per foot (30 cm) of row. If we use organic materials, such as composted manure, the same rate of nutrient would be provided by a layer of compost no more than ⅛ inch (3 mm) thick. A ½-inch (1.3 cm) layer would provide the equivalent of 360 pounds per acre (400 kg/ha) of each nutrient.

These examples assume that all the nutrients are applied in a single application. If you are using bonemeal or a high-phosphorus mineral fertilizer as a starter for your transplants, cut back on the phosphorus in the additional fertilizer. Problems with nutrient imbalances and salt injury most often arise when multiple nutrient sources are used without taking into account the overall impact.

Among the mineral fertilizers, there is a division between "regular" and "premium" products. The biggest difference in the premium blend (aside from price) is the addition of micronutrients. The potential benefit of the micronutrients depends on whether your soil is deficient in one or more of those elements and whether the plant you're growing needs more of those elements than the soil can supply. Micronutrient additions work well for potted plants, since typical potting soil has almost no nutrients. Potting mixes also have a very low capacity to bind nutrients, so what is added at the surface is able to get down to the roots. For plants grown in soil, however, the justification for these micronutrient blends is less clear.

Where I live, the glacial soils are geologically very young and contain large reserves of most micronutrients. The soil should be able to supply what the plants need, in the same way that eating a varied diet should provide all the vitamins and minerals we need. In this case, the micronutrient package is comparable to a daily multivitamin. It is something we take just to be sure we never run short, rather than because we need it. For gardens on the coastal plain in Georgia or South Carolina, the situation is quite different. The highly weathered soils have much lower quantities of micronutrients and are, therefore, more likely to benefit from the added minerals.

Adding Nutrients

There are advantages and drawbacks to both organic and mineral sources of nutrients, some of which are summarized in the following table.

Mineral Fertilizers	Organic Materials
Quick and consistent availability.	Sustained availability as materials decompose; rate of decomposition affected by weather.
Available in multiple formulations with known nutrient contents, so applications can be tailored to meet plant requirements.	Limited range of formulations; proportions of nutrients in material may not match plant needs.
Concentrated, so relatively little product to handle.	Nutrient content more diluted, so relatively large volumes of material must be handled.
High risk of salt injury from some products if overapplied.	Low but not negligible risk of salt injury.
Adds only nutrients to soil.	Adds significant quantities of organic matter along with nutrients.

Given the rapid availability of nutrients from mineral fertilizers, expect a quick response in plant growth. But there is the risk that a mobile nutrient like nitrogen may literally "run out" before the plant finishes growing. The slow release from organic materials, on the other hand, provides a sustained supply of nutrients, which is an advantage for plants that require nutrients over the entire growing season. Organic materials may not be the best choice for short-season crops, however, particularly ones that are planted early into cool soil, as nitrogen may not be released fast enough to meet the plant's needs.

Another potential drawback to organic materials is that nutrients may continue to be released after the plants no longer need them. High levels of nitrogen can remain in the soil at the end of the growing season, raising environmental concerns if they leach into the groundwater and causing problems with woody plants that don't harden off properly. Including cover crops in the garden rotation can alleviate both concerns.

How Much Product Should You Apply?

A soil test report for your garden may include a recommendation expressed as the amount of a particular product you need to apply. Simply go to the garden centre and ask for the appropriate amount of whatever fertilizer blend is recommended. In other cases, the recommendation will be expressed as the amount of nutrient you need to apply (N, P_2O_5 or K_2O) per 1,000 square feet or per 100 square metres (which are very close to the same area). You need to do a little bit of calculation to figure out how much fertilizer to apply.

To convert the amount of nutrient to apply into the quantity of fertilizer, you must know the concentration of the nutrient in the fertilizer. Once you know this, convert the percentage of nutrient (from the fertilizer label) into a decimal by dividing by 100. For example, 10-10-10 fertilizer contains 10 percent each of nitrogen, phosphate and potash, so the decimal equivalent is 0.1 for each nutrient. Using your calculator, divide the amount of nutrient recommended for your garden by the decimal, and the result is the amount of fertilizer you need to apply to the area listed for the recommendation. Then convert this to the amount of fertilizer you need to apply by multiplying by the area of your garden (i.e. how many 1,000 square feet or 100 square feet of total garden area you have).

The balance of nutrients in an organic material like compost or manure is dictated by the raw materials that went into it and how many nutrients have been lost during processing and storage. No organic material can provide a single nutrient to address a specific deficiency. Fortunately, most plants are quite tolerant of the extra nutrients in organic materials.

There is a common misconception that salt injury is limited to the application of mineral fertilizers, but high rates of manure or compost can also raise the concentration of salt in the soil to harmful levels. The nutrients most often associated with salt injury are nitrogen (particularly in the ammonium form) and potassium, especially when they are applied close to seeds or transplants. Phosphate is generally not a concern for salt injury, since it reacts so quickly with soil minerals that it never generates a high concentration in the soil solution.

When deciding whether to add an organic or a mineral nutrient source to your garden, keep in mind the difference between nutrients that are mobile in the soil, like nitrogen and sulfur, and nutrients that are held in the soil, such as phosphorus and potassium. Mobile nutrients can leach out of the soil whenever the amount of precipitation exceeds the rate of evapotranspiration, such as during the period from late fall to early spring throughout much of eastern North America. As a result, very little nitrogen and sulfur are retained from one growing season to the next, and these nutrients must be added to the soil each year for the plants that require them (legumes fix their own nitrogen and don't need any addition). Nutrients that are held tightly to soil particles build up over time if more are added than are taken away in the harvested crops. Many gardens have been built up to the point that there is enough phosphorus and potassium to supply the plants for decades to come, and adding more could be harmful to plant growth.

Interactions Between Nutrients

When chemist Justus von Liebig formulated the "law of the minimum" in the early 19th century, he ignored the possibility of interactions between nutrients by stating that the nutrient with the least supply was what limited plant growth. We now know that in some circumstances, the interactions between nutrients can have a big influence on plant nutrition. These interactions, in general, become important only when there is a major imbalance between the quantities of the interacting nutrients, which is exactly the condition we can create with liberal applications of fertilizer and compost. A few of the interactions you may see in your garden are described here.

Nitrogen to Sulfur

An inadequate supply of sulfur limits the plant's production of protein, which is where nitrogen is used in the plant. If you increase the supply of nitrogen to a sulfur-deprived plant, that nitrogen has nowhere to go. This imbalance worsens the plant's symptoms of sulfur deficiency and inhibits its growth. Adding sulfate sulfur (potassium sulfate or gypsum) should alleviate this imbalance. In future, reduce the application of nitrogen.

Nitrogen to Potassium

Specific to the ammonium form of nitrogen, this interaction has two impacts on potassium availability. The first is competition for uptake by the roots, resulting in excess ammonium pushing potassium out of the way and leaving the plant with a deficiency. The second impact is on the supply of available potassium from clay soils. The same mechanism that fixes potassium between clay layers can also trap ammonium. A sudden influx of ammonium into the system can trap potassium behind it, delaying the release of the potassium ions into the soil solution. Fortunately, both of these impacts are temporary, as the ammonium is soon converted to the nitrate form.

Phosphorus to Zinc

Adding phosphorus fertilizer to a soil already high in phosphorus can induce a zinc deficiency (stunted growth, with pale or white patches at the base of the new leaves). Adding zinc doesn't alleviate the symptoms, but the presence of the zinc deficiency alerts you to stop applying phosphorus.

Potassium to Magnesium

High levels of potassium can interfere with magnesium uptake. Plants growing in soil that normally has adequate magnesium begin showing deficiency symptoms (pale-colored striping between the veins or reddish or purple discoloration on the back of the leaves). This interaction does not work the other way, so there is no concern with high levels of magnesium interfering with potassium uptake.

Micronutrient Imbalances

Most micronutrients are metals, and they all enter the root through similar pathways. An excess of one interferes with the uptake or the function of the others within the plant. This is usually an issue with soil contamination from industrial sources, but it can also be the result of applying a micronutrient fertilizer where it isn't needed.

Summing Up: The Principles of Managing Nutrients

Nutrient behavior in soil is complex. Fortunately, the buffering capacity of soils provides us with a lot of management flexibility, so we do not have to hit a precise level of nutrient concentration to grow healthy plants. Managing nutrients to keep your garden growing well can be boiled down to three fairly simple rules:

1. Maintain good soil tilth and organic-matter levels.
2. Return as many nutrients to the soil as possible using what has been grown in your garden.
3. Apply fertilizer or compost when needed, but not too much.

For More Information

Soil Fertility Handbook. Ontario Ministry of Agriculture, Food and Rural Affairs, 2006. Publication 611. Queen's Printer for Ontario, Toronto, Canada. This book covers many of the aspects of soil fertility management in more detail, but from an agricultural perspective.

Sustaining the Land

Whoever could make two ears of corn, or two blades of grass, to grow upon a spot of ground where only one grew before, would deserve better of mankind, and do more essential service to his country, than the whole race of politicians put together.

— Jonathan Swift, *Gulliver's Travels*

13

WE TEND TO think about sustainable land management as something that should concern farmers, who have extensive tracts of land under their control, rather than something backyard gardeners need to worry about. But gardeners, too, must practice good stewardship of the soil. In fact, lack of care by gardeners can be just as harmful to the environment as irresponsible large-scale farming practices.

You may wonder why the impact is so great, since it is not readily visible. Gardening is an increasingly popular activity among householders, and even though the impact of an individual garden may be relatively small, the cumulative effect quickly adds up. Since the management of most garden plots is quite intensive, individual gardeners make more additions to their gardens per square foot than farmers do to their fields. In addition, there is a greater opportunity for nutrients and other essential compounds to move from the garden: The transport pathway from the garden to surface water, particularly in urban areas, is much shorter than it is from most agricultural fields to the nearest stream. Storm sewers are extremely effective at carrying runoff—and dissolved nutrients—from our gardens (and from streets and sidewalks) to the nearest waterway. This increased ease of transport substantially reduces the land's capacity to trap nutrients and reuse them.

Gardeners are often guilty of squandering nonrenewable resources. Fertilizer, which is either produced from fossil fuels or mined from finite reserves of nutrient-bearing minerals, is one obvious example. But it is not the only one. The same gardener who is troubled about modern farming practices that rely on purchased inputs may happily haul leaves and dead plants to the curb, then head off to the garden center to buy peat moss to replace the organic matter that has just been removed from the garden. Not only does peat moss provide lower nutrient levels than the materials leaving the yard, but there is a heated debate about how "renewable" the peat moss resource is and whether the extraction of sphagnum peat should be considered mining rather than harvesting.

I believe that every gardener wants to be a good environmental steward. It is impossible to spend time working with soil and plants without developing an awareness of our connection with the natural world. The problem is that any adverse effects from the way we garden are removed from where we live, and that connection is not always easy to see. I hope to open your eyes to this connection with nature and to encourage you to garden in a way that is more in tune with the global cycles of water, energy and nutrients. A garden can be beneficial or harmful to the environment, and the way we manage our gardens makes all the difference.

Preventing Soil Erosion

Erosion is perpetually reshaping the landscape, wearing down mountains and creating the sediment that forms most of our soils. It's a natural process. Too much erosion, however, carries away productive soils faster than they are created. When eroded sediment reaches our streams and rivers, it can affect water clarity, making it unsuitable for many species of fish and invertebrates and potentially fouling the spawning beds for sport fish. As it moves downstream, the sediment accumulates in harbors, where costly dredging may be required to keep them open to boat traffic.

The problem, then, is twofold. Erosion affects the productivity of our gardens and damages our streams and rivers. Remedies can take many forms, depending on the situation. Ideally, strategies for preventing soil erosion can be incorporated into your landscape design in a way that enhances its beauty.

Most of these strategies are exactly the same as those for maintaining good tilth and an active soil biology. A stable soil structure allows water to infiltrate the soil rather than run off the surface. Less runoff means less soil is carried away, and the infiltrating water redeposits any sediment it is carrying. A stable soil structure also resists detachment by wind.

Continuous Plant Cover

Wind and water attack exposed soil, breaking particles loose and carrying them away. Keeping an umbrella of living plants over the soil protects it from these ravages. The leaves absorb the energy of raindrop impact, allowing the moisture to filter gently down to the soil surface. The stems slow the wind, so there is less opportunity for it to scour the surface of the soil. The roots of the growing plants form a lattice to hold the soil together, so it can better resist detachment. (The ultimate example of continuous plant cover is turf, or grass, because the plants grow very close together and are present year-round, effectively preventing soil erosion.)

In the garden, we need to modify the approach from "continuous" to "as close to continuous as we can manage." This means avoiding long periods of fallow ground by planting another crop as soon as one plant matures and is harvested. It may be a second crop after an early crop, such as beans following lettuce or radishes, or it may be a cover crop that will protect the ground until the next season.

Plant Residue

If you leave the remains of harvested plants on the surface of the soil, they perform the same protective role as living plant material. Plant residue does two things: It absorbs the impact of raindrops (as do living plants), and it creates little dams that slow water as it runs over the surface, allowing it time to either infiltrate the soil or drop any sediment it is carrying.

The rule of thumb is that about one-third of the soil surface should be covered with plant residue to protect it from wind or water erosion. That isn't too difficult to achieve with a crop like sweet corn or peas, where the plants are fairly large relative to the harvested portion. But if you are growing lettuce or carrots, the plant residue left after harvest won't provide adequate cover. Another challenge is that the residue of many vegetable crops breaks down very quickly, leaving the surface unprotected.

Even weeds can play a role as a cover crop or can provide cover if you pull them out and leave them on the soil surface. The obvious caution is that weeds that have gone to seed must be removed from the garden rather than left to increase the weed population. You will likely find, however, that the weeds pulled early enough to prevent seed set are pretty tender, so the cover they provide doesn't last very long.

Mulch

One advantage we backyard gardeners have over farmers is that our plots are small enough to spread mulch over the entire surface. Mulching holds soil moisture and suppresses weed growth, and it also protects the soil from erosion by wind or water. (If, during a heavy rain, enough water flows over the surface to physically shift the mulch, it's time to reshape your garden so that erosion does not become a problem.)

Water Diversion

Water that runs off the garden may be less of a problem than water that runs into the garden from higher ground. That water should be diverted away from your garden to an alternative pathway where it won't do any harm—perhaps into a grassed channel beside the garden or a subsurface drain that will carry excess water below the garden. The grassed waterway should have a flat bottom and gently curved sides; otherwise, a gully may form down the center. If there is more water than can be comfortably carried away by a grassed waterway, consider consulting an engineer for advice on building something that can withstand the erosive force of the water without creating problems elsewhere.

Terraces

Many human civilizations have successfully farmed on steeply sloping ground by creating terraces—flat areas separated by slopes stabilized with permanent vegetation or by retaining walls built with wood, brick or stone. Costly to install, terraces may not accommodate some of your garden machines, such as rototillers, but when properly built, they essentially halt soil erosion. If you have a city lot on a steep slope, it may be the only way to have gardens. Terraces also create a dramatic visual impact.

Windbreaks

In some areas, wind erosion is as much of a challenge as water erosion. A windbreak can make a big difference in the amount of soil loss by wind. It also moderates the temperature and moisture loss near the windbreak. Its effect extends about 20 times the height of the windbreak downwind and two to five times its height upwind. A semiporous tree windbreak is more effective than a solid wall, and the trees slow runoff and reduce soil erosion.

What Happens in the Garden Should Stay in the Garden

To manage nutrients properly, don't apply them at rates that will overwhelm the capacity of the plants to use them or the soil to hold them. When too many nutrients are applied, either they are released into surface runoff or they leach into groundwater.

A colleague in a U.S. state where nutrient runoff from intensive livestock agriculture is a serious concern requested a summary of the top 400 samples from the state soil testing lab. He was surprised to learn that half of the samples with the highest phosphorus levels were not from livestock farms. Instead, they came from urban and suburban lawns and gardens, including the sample that reported the highest level of phosphorus accumulation in the soil. This is typical of the results found in many states and provinces.

High phosphorus in surface runoff causes excess algae growth in lakes and rivers. Harmful to fish and other aquatic life because they use up oxygen when they die and decompose, algae also impart an unpleasant taste and odor to the water that conventional water treatment cannot remove. In some cases, algae blooms release toxic chemicals into the water, making it unsafe to drink or to bathe or swim in. Since the phosphate ion is so reactive, much of the phosphorus that reaches surface water is attached to soil particles that have eroded. There are limits to the amount of phosphorus the soil can hold, however, so at high levels, the phosphate ions are not held as tightly and can be dissolved in the runoff water. This dissolved phosphorus is available to aquatic algae and has a much greater impact than a similar amount of phosphorus bound to soil particles. Garden soils dominate the group of soils that have high phosphorus concentrations and lose significant quantities of dissolved phosphorus.

Ammonium nitrogen losses from soils are a concern because ammonia is highly toxic to fish. This usually poses a problem only if there is runoff immediately after the application of raw manure or an ammonium-based fertilizer, such as urea. Nitrate loss from gardens is more common than ammonium loss. Elevated nitrate levels in surface water have been associated with birth defects in some amphibians. If high nitrate in groundwater exceeds the drinking-water limit of 10 milligrams of nitrate-nitrogen per liter, it can lead to municipal wells being shut down, since there is no practical and affordable way to remove the nitrate.

Avoid off-site impacts from your gardening activities by limiting the amount of nutrients you add. Whether you use a mineral or an organic fertilizer, the quantity you apply is the key issue—and too much is too much. Once you've built up your soil, you need to apply only as much nutrient as the plants remove each year. On the scale of a garden, this is not very much. One plant with a big appetite for nutrients is the tomato, but an excellent crop of tomatoes on a 1,000-square-foot (93 sq m) garden removes about 2 pounds (1 kg) of phosphate and 10 pounds (4.5 kg) of potash, assuming all the tomato vines are removed along with the fruit. Most other vegetable crops use less than half this amount, so if you apply more than 5 pounds

(2.2 kg) of 18-18-18 fertilizer, you're applying more phosphate than is being removed.

Soils and Climate Change

The gases emitted from soil profoundly influence our atmosphere, and the way we manage our soil significantly impacts how much of those gases is released.

The greenhouse effect is caused when gases in the atmosphere that are transparent to visible light reflect infrared rays back to Earth. In this way, the atmosphere acts like the glass in a greenhouse, keeping the temperature inside the greenhouse warmer than it would otherwise be. It has been estimated that without this effect, Earth would be about 90 Fahrenheit degrees (50 Celsius degrees) cooler and life as we know it would not be possible.

The highest concentrations of greenhouse gases are carbon dioxide, methane and nitrous oxide. All three are associated with processes that occur in the soil. Their concentrations have been increasing since the beginning of the industrial revolution, and they are clearly having a major impact on our climate. We should, therefore, be trying to limit emissions as much as possible. Gardeners can make a real difference in this regard.

Organic matter in the soil is a huge store of carbon, and as the carbon decomposes, it is released as carbon dioxide. Plants use this carbon dioxide to grow, and if the rate of carbon fixed by plants equals or exceeds the rate of decomposition, there is no net release of carbon dioxide to the atmosphere from the soil. When early settlers to this continent cleared large areas of forests and the prairies were first plowed, there was a burst of organic-matter decomposition, which pumped a lot of carbon dioxide into the atmosphere. Most farming systems today are near equilibrium, so gains and losses of carbon dioxide are in balance. But how do backyard gardens compare?

Unfortunately, not very well. We build the organic matter of our soil by using external sources of organic matter rather than relying on photosynthesis to fix carbon in place. Even worse, we often add far more organic matter than the soil can hold, so the rate of decomposition is much faster than it is in most agricultural soils. And many of us till the ground frequently and intensively, which further hastens the release of carbon dioxide.

About half of the methane in the atmosphere is released from soils, primarily from flooded soils and wetlands. Unless you are raising paddy rice, you are unlikely to be a significant contributor of methane losses to the atmosphere.

The biggest impact of agriculture on greenhouse-gas emissions is from nitrous oxide, a product of denitrification in the soil. The rate of denitrification is determined by three factors: a lack of oxygen in the soil due to excess water, a food supply for the microbes and a supply of nitrate. The water supply varies with the weather, so we can't control that, apart from ensuring there is good drainage to prevent the water from overstaying. However, we have a lot of influence over the microbe food supply and the nitrate supply. And, once again, we create conditions in the garden that result in more nitrous oxide emissions than in most agricultural fields. The food source for microbes is the easily degraded carbon compounds in the organic amendments we add to garden soils. Under aerobic conditions, this food is a good thing, but not so much if a heavy rainfall saturates the soil and creates anaerobic conditions. This is when the supply of nitrate becomes the critical factor and when our tendency to apply fertilizer or compost too liberally becomes problematic.

Mitigating the carbon dioxide and nitrous oxide emissions from garden soils returns us to a familiar theme: moderation. Add organic materials, but not too much. Add nitrogen, but not too much. Work the soil to dry it out, but not too much.

Growing organic matter in place with cover and forage crops in rotation is more sustainable than bringing in external sources of organic matter. Not only do these crops tie up atmospheric carbon dioxide while they are growing, but they soak up nitrate from the soil so that it is not available to the denitrifying bacteria. If you want to add organic materials, apply them as mulch. This will have less impact on the microbial community than mixing them into the soil, because the contact area with the soil is limited to the boundary between the soil and the mulch.

Protecting the Soil Resource

There is a saying in many cultures that goes something like this: "We don't own the soil; we simply borrow it from our grandchildren." We have a responsibility to protect and preserve the soil and, as much as possible, to leave it in better condition than we received it. By gardening, we have an opportunity not only to create beauty and bounty with the plants we grow but to build a fertile, resilient soil that will sustain our families—and the families that follow—for years to come.

And now the soil is in your hands. One of the lessons I hope you've embraced in reading this book is that soils vary widely and that the way you manage them must be specific to your own soil conditions. There is no one "recipe" that can solve all soil problems. Any general recommendations are wrong far more often than they are right.

If you take away the following key concepts, you've gone a long way toward becoming a better steward of your soil:

Manage for the Soil You Have and Not for Some Ideal Soil

While every soil is unique, soils that have developed in the same area usually share a number of characteristics. When you understand your soil, you can begin to sort out what advice makes sense for your soil and to make decisions that minimize its weaknesses and accentuate its advantages.

Everything in Moderation

Fertilizer, water, organic amendments—all play a role in successful gardening, but too much of any one of them can hurt your plants and potentially damage your garden for years to come. Resist the temptation to add "just a little bit extra."

What's Good for Your Soil Is Good for Your Plants

Adequate moisture and air, moderate temperatures and an ample nutrient supply benefit both the microorganisms living in your soil and your plants' roots. Establish these conditions, and your garden will develop an active soil biology, including improved soil structure, better resistance to compaction and surface crusting, suppression of plant diseases and a rapid release of nutrients from plant residues in the soil.

Proper management of your soil requires effort, both mental and physical, but the end result is healthier, more vigorous plants. In the long run, it means less work as your soil becomes more forgiving and—the greatest reward of all—you get to sit back for a few moments and enjoy your garden.

Happy gardening!

For More Information

The beginning of the "land ethic" can be traced to *A Sand County Almanac* by Aldo Leopold (Oxford University Press, 1949), which is widely available in reprint. It is recommended reading for anyone with an interest in conservation and land management.

Two books that provide a broader historical context for the importance of good land management are:

- *Dirt: The Erosion of Civilizations* by David R. Montgomery. University of California Press, 2007.
- *Collapse: How Societies Choose to Fail or Succeed* by Jared Diamond. New York: Penguin Books, 2005.

Glossary of Soil Terms

anion An ion with a negative electrical charge. Examples of common anions in soil are nitrate, phosphate, sulfate and chloride.

calcareous soil A soil that contains calcium carbonate as one of its constituents; fizzes when dilute hydrochloric acid is dropped on it.

cation An ion with a positive electrical charge. Examples of common cations in the soil are calcium, magnesium, potassium and ammonium.

CEC (cation exchange capacity) The capacity of the soil to hold on to positively charged ions (cations) and to exchange them for other cations in the soil solution.

clay The size of the mineral particles in the soil are less than 0.002 millimeters in diameter; feels sticky when moist.

field capacity The moisture content of the soil when the large pores have drained and the remaining water is held more tightly than the force of gravity.

horizon A term used by soil scientists to describe the horizontal layers that constitute a complete soil profile; it is designated by a capital letter (A, B, C), with lowercase modifiers that provide additional information about that horizon. The A horizon is closest to the surface (topsoil), and the C horizon is the parent material from which the soil formed.

humus Decayed organic matter that has formed stable organic compounds in the soil which are resistant to further breakdown.

pH A measure of the acidity or alkalinity in a solution, on a scale from 0 to 14. Neutral pH (neither acidic nor alkaline) is 7. Values below 7 indicate acid conditions, while values above 7 indicate alkaline conditions.

physiography The study of the physical features, or landforms, on the Earth's surface.

sand Mineral particles in the soil that range from 0.05 to 2 millimeters in diameter; feels gritty when rubbed between fingers.

silt The mineral particles in this soil range from 0.002 to 0.05 millimeters in diameter; feels floury when rubbed between fingers.

structure The way in which individual soil particles are organized into larger units, forming crumbs, clods, plates, blocks or prisms.

surficial geology The study of landforms and the unconsolidated sediments that lie beneath them.

texture The proportion of sand, silt and clay in the mineral part of the soil.

tilth The physical condition of the soil with respect to its fitness for cultivation and planting.

water table The top of a saturated layer in the soil. If you dig a hole below the water table, water will run into the hole and fill it to the level of the water table.

wilting point The moisture content of the soil when plants can no longer extract water to meet their needs (and hence wilt) because the remaining water is held too tightly in fine pores or to the surface of soil particles.

Appendix I:
Selected Laboratories for Testing Garden Soils

THE MOST RELIABLE source of information on soil test labs is your local agricultural or horticultural extension office (see Appendix II). There, you can find out which labs perform the appropriate tests for your area. Up-to-date listings of labs suitable for testing garden soils in your area are also available from the North American Proficiency Testing Program (**www.naptprogram.org/about/participants/all**). Check with the lab directly for prices and availability of specific analyses.

The following list is not exhaustive, nor is it intended as an endorsement of any particular lab, but it will serve as a starting point.

Canada

British Columbia

AGAT Laboratories/Burnaby
Burnaby, BC V5J 0B6
Tel.: (778) 452-4000
krusberski@agatlabs.com
www.agatlabs.com

ALS Environmental/Vancouver
Burnaby, BC V5A 1W9
Tel.: (604) 253-4188
quality.vancouver@alsenviro.com
www.alsglobal.com

British Columbia Ministry of Environment Analytical
Chemistry Laboratory
Victoria, BC V8Z 5J3
Tel.: (250) 952-4133
clive.dawson@gov.bc.ca

Prairie Provinces

ALS Laboratory Group/Edmonton
Edmonton, AB T6E 0P5
Tel.: (780) 413-5988
anne.beaubien@alsglobal.com
www.alsenviro.com

Farmer's Edge Laboratories
Winnipeg, MB R2J 0H3
Tel.: (204) 233-4099, ext. 462
Patrick.Visser@farmersedge.ca
www.farmersedge.ca

Ontario

A&L Canada Laboratories, Inc.
London, ON N5V 3P5
Tel.: (519) 457-2575/Toll-free: (855) 837-8347
dstallard@alcanada.com
www.alcanada.com

SGS Agri-Food Laboratories
Guelph, ON N1H 6T9
Tel.: (519) 837-1600/Toll-free: (800) 265-7175
papken.bedirian@sgs.com
www.agtest.com

Stratford Agri Analysis
Stratford, ON N5A 6W1
Tel: (519) 273-4411
Toll-free: (800) 323-9089, ext. 30
tbeaucage@masterfeeds.com
www.stratfordagri.ca

University of Guelph Laboratory Services
Guelph, ON N1G 2W6
Tel.: (519) 767-6299/Toll-free: (877) 863-4235
nschrier@uoguelph.ca
www.uoguelph.ca/labserv

Quebec

GeoSoil Laboratory
Mont-Saint-Hilaire, QC J3G 4S6
Tel.: (450) 464-2522
pierre.lamoureux@synagri.ca

Laboratoire d'analyse SM
Longueuil, QC J4G 1P1
Tel.: (450) 674-5271
ddesbiens@groupesm.com

Atlantic Provinces

New Brunswick Agricultural Lab
Fredericton, NB E3B 8B7
Tel.: (506) 453-3495
alan.scott@gnb.ca

NL Provincial Soil, Plant and Feed Lab
St. John's, NL A1B 4J6
Tel.: (709) 729-6738
youjiao@gov.nl.ca

PEI Analytical Laboratories
Charlottetown, PE C1A 4N6
Tel.: (902) 368-5622
mcmacneill@gov.pe.ca

United States

Northeast

A&L Eastern Labs, Inc.
Richmond, VA 23237
Tel.: (804) 743-9401
pmcgroary@aleastern.com
www.aleastern.com

Ag Analytical Services Lab
University Park, PA 16802
Tel.: (814) 863-0841
pxs32@psu.edu
www.aasl.psu.edu

Agri Analysis, Inc.
Leola, PA 17546
Tel.: (717) 656-9326/Toll-free: (800) 464-6019
susanm@agrianalysis.com
www.agrianalysis.com

Cornell Nutrient Analysis Laboratory
Ithaca, NY 14853
Tel.: (607) 255-4540
td47@cornell.edu
www.cnal.cals.cornell.edu

Rutgers Soil Testing Laboratory
New Brunswick, NJ 08901
Tel.: (848) 932-9295
soiltest@njaes.rutgers.edu

University of Delaware Soil Testing Program
Newark, DE 19716-2170
Tel.: (302) 831-1392
kgartley@udel.edu
http://ag.udel.edu/dstp

University of Maine Analytical Lab
Orono, ME 04469-5722
Tel.: (207) 581-3591
hoskins@maine.edu
http://anlab.umesci.maine.edu

University of Massachusetts Amherst Soil and Plant Tissue Testing Lab
Amherst, MA 01003
Tel.: (413) 545-2311
spargo@umext.umass.edu

Southeast

A&L Analytical Laboratories, Inc.
Memphis, TN 38133
Tel.: (901) 213-2400/Toll-free: (800) 264-4522
clangford@allabs.com
www.allabs.com

Agriculture Diagnostic Service Laboratory
University of Arkansas
Fayetteville, AR 72704
Tel.: (479) 575-3908
nwolf@uark.edu

Auburn University Soil Testing Laboratory
Auburn University, AL 36849-5411
Tel.: (334) 844-3958
hulukgo@auburn.edu
www.aces.edu/soiltest

North Carolina Agronomic Services/
Soil Testing
Raleigh, NC 27699-1040
Tel.: (919) 733-2655
david.hardy@ncagr.gov

Soluciones Analiticas
Doral, FL 33172-5028
Tel.: (305) 594-7675
calidad@solucionesanaliticas.com
www.solucionesanaliticas.com

Southeastern Agricultural Lab, Inc.
Barney, GA 31625
Tel.: (229) 775-2426
seagrilab@msn.com
http://seaglab.com

University of Arkansas Soil Testing Lab
Marianna, AR 72360
Tel.: (870) 295-2851
dcarroll@uark.edu

University of Florida Analytical Research Lab
Gainesville, FL 32611-0740
Tel.: (352) 392-1950, ext. 221
nancysw@ufl.edu

North-Central

A&L Great Lakes Laboratory
Fort Wayne, IN 46808-4414
Tel.: (260) 483-4759
Fax: (260) 483-5274
bthayer@algreatlakes.com
www.algreatlakes.com

AgSource/Bonduel
Bonduel, WI 54107
Tel.: (715) 758-2178
speterson@agsource.com
www.agsource.com

AGVISE Laboratories/Benson
Benson, MN 56215
Tel.: (320) 843-4109
cindye@agvise.com
www.agvise.com

Iowa State University Soil and Plant Analysis Laboratory
Ames, IA 50011-1010
Tel.: (515) 294-3076
soiltest@iastate.edu
www.agron.iastate.edu/soiltesting

MN Valley Testing Laboratories, Inc./
New Ulm
New Ulm, MN 56073-0249
Tel.: (507) 233-7139
mbaumgart@mvtl.com
www.mvtl.com

Ohio State University
Ohio Agricultural Research and Development Center
Service Testing and Research Lab
(STAR Lab OSU/OARDC)
Wooster, OH 44691
Tel.: (330) 263-3683
jewell.4@osu.edu

Perry Agricultural Laboratory
Bowling Green, MO 63334
Tel.: (573) 324-2931
bob@perryaglab.com

SGS North America
Belleville, IL 62221
Tel.: (618) 233-0445
marina.pantu@sgs.com
www.alveylabs.com

Spectrum Analytic
Washington Court House, OH 43160
Tel.: (740) 335-1562
vernon@spectrumanalytic.com
www.spectrumanalytic.com

Sure-Tech Laboratories
Indianapolis, IN 46221
Tel.: (317) 243-1502
jmjaynes@landolakes.com

United Soils, Inc.
Fairbury, IL 61739
Tel.: (815) 692-2626
ardeleanc@unitedsoilsinc.com
www.unitedsoilsinc.com

University of Missouri Soil and Plant Testing Laboratory
Columbia, MO 65211
Tel.: (573) 882-3250
nathanm@missouri.edu
http://soilplantlab.missouri.edu/soil

Great Plains

AgLab Express
Sioux Falls, SD 57105-6411
Tel.: (605) 271-9237
anthonybly@aglabexpress.com
http://aglabexpress.com
AgSource/Harris
Lincoln, NE 68502
Tel.: (402) 476-0300
lclaassen@agsource.com
www.agsource.com

AGVISE Laboratories/Northwood
Northwood, ND 58267
Tel.: (701) 587-6010
lwikoff@polarcomm.com
www.agvise.com

Brigham Young University
Provo, UT 84602
Tel.: (801) 422-2147
rachelbuck@byu.edu

Kansas State University Soil Testing Laboratory
Manhattan, KS 66506
Tel.: (785) 532-7897
dmengel@ksu.edu
www.ksu.edu/agronomy/soiltesting

Midwest Laboratories
Omaha, NE 68144-3617
Tel.: (402) 334-7770
djabs@midwestlabs.com
www.midwestlabs.com

Olsen's Agricultural Laboratory
McCook, NE 69001-0370
Tel.: (308) 345-3670
kevin@olsenlab.com
www.olsenlab.com

Platte Valley Laboratories
Gibbon, NE 68840
Tel.: (308) 468-5975
stu@soillab.com
www.soillab.com

Servi-Tech Laboratories/Hastings
Hastings, NE 68902
Tel.: (402) 463-3522
hansb@servitechlabs.com
www.servitechlabs.com

SGS Mid-West Seed Services, Inc.
Brookings, SD 57006
Tel.: (605) 692-7611, ext. 5
christina.sternhagen@sgs.com
www.cropservices.sgs.com

Ward Laboratories, Inc.
Kearney, NE 68847
Tel.: (308) 234-2418
rayward@wardlab.com
www.wardlab.com

Northwest

A.V. Labs, Inc.
Soil and Plant Program
Othello, WA 99344
Tel.: (509) 488-2468
von@avlabsinc.com

Cascade Analytical
Wenatchee, WA 98801
Tel.: (509) 662-1888
briannap@cascadeanalytical.com
www.cascadeanalytical.com

Central Analytical Lab
Corvallis, OR 97331
Tel.: (541) 737-5731
will.austin@oregonstate.edu
http://cropandsoil.oregonstate.edu/cal

Northwest Agricultural Consultants
Kennewick, WA 99336
Tel.: (509) 783-7450
bob@nwag.com
www.nwag.com

SoilTest Farm Consultants
Moses Lake, WA 98837
Tel.: (509) 765-1622
brent@soiltestlab.com
www.soiltestlab.com

Stukenholtz Laboratory, Inc.
Twin Falls, ID 83301
Tel.: (208) 734-3050
lab@stukenholtz.com
http://stukenholtz.com

USAg Analytical Services Inc.
Pasco, WA 99301-4244
Tel.: (509) 547-3838
usag.inc@qwestoffice.net

Western Laboratories
Parma, ID 83660
Tel.: (208) 722-6564
Cathy@westernlaboratories.com
www.westernlaboratories.com

Southwest

A&L Western Agricultural Labs
Modesto, CA 95351
Tel.: (209) 529-4080
rbutterf@al-labs-west.com
www.al-labs-west.com

Dellavalle Laboratory Inc.
Fresno, CA 93728-1221
Tel.: (559) 233-6129
pmiller@dellavallelab.com
www.dellavallelab.com

Denele Agrilink Laboratories
Turlock, CA 95380
Tel.: (209) 634-9055
Fax: (209) 634-9057
denelelab@sbcglobal.net

Inter Ag Services Laboratory
Phoenix, AZ 85034-6912
Tel.: (602) 273-7248
sheri@iaslabs.com
www.iaslabs.com

Laguna Environmental Lab
Santa Rosa, CA 95407
Tel.: (707) 543-3363
ckaul@srcity.org
www.santarosautilities.org

Precision Agri-Lab
Madera, CA 93637
Tel.: (559) 661-6386
chad.bethel@cpsagu.com

Servi-Tech Laboratories/Amarillo
Amarillo, TX 79109
Tel.: (806) 677-0093
toddw@servi-techinc.com
www.servitechlabs.com

SFASU Soil Plant and Water Analysis Laboratory
Nacogdoches, TX 75962-9020
Tel.: (936) 468-4500
wweatherford@sfasu.edu
http://soils.sfasu.edu

Sunland Analytical Lab, Inc.
Rancho Cordova, CA 95670
Tel.: (916) 852-8557
rhsunland@sbcglobal.net

UC Davis Analytical Laboratory
Davis, CA 95616
Tel.: (530) 752-0147
anlab@ucdavis.edu
http://anlab.ucdavis.edu

USU Analytical Lab
Logan, UT 84322-9400
Tel.: (435) 797-2217
usual@usu.edu
www.usual.usu.edu

Valley Tech Agricultural Lab
Tulare, CA 93274
Tel.: (559) 688-5684
sam@vtaglab.com
www.vtaglab.com

Appendix II: Agricultural and Horticultural Extension Resources

United States

Information about agriculture and horticulture in the United States is available through a network of Cooperative Extension System Offices operated by the U.S. Department of Agriculture. For a list of offices in each state, go to **www.csrees.usda.gov/Extension/index.html.**

You can also obtain information directly from the USDA eXtension site (**www.extension.org**), including links to the Master Gardeners in each American state and Canadian province. Master Gardeners are excellent resources regarding most gardening issues and have detailed local knowledge about adapted plant species and common pests.

Finally, a couple of resources that I have found to be useful and refreshing in their approach are the Horticultural Myths website by Linda Chalker-Scott at Washington State University (**http://puyallup.wsu.edu/~linda%20chalker-scott/Horticultural%20Myths_files/index.html**) and The Garden Professors blog, to which Chalker-Scott contributes (**https://sharepoint.cahnrs.wsu.edu/blogs/urbanhort/default.aspx**).

Canada

Information about agriculture and horticulture in Canada (including home gardens) is provided through the provincial agriculture departments. Contact information for each of the provincial offices is provided opposite.

British Columbia

British Columbia Ministry of Agriculture
Victoria, BC V8W 9E2
Tel.: (250) 387-5121
www.gov.bc.ca/agri

Prairie Provinces

Alberta Agriculture and Rural Development
Edmonton, AB T6H 5T6
Toll-free: 310-FARM/310-3276 (Alberta only)
Tel.: (403) 742-7901 (outside Alberta)
www.agric.gov.ab.ca/app21/ministrypage

Manitoba Agriculture, Food and Rural Initiatives
www.gov.mb.ca/agriculture/index.html

Saskatchewan Agriculture
www.agriculture.gov.sk.ca

Ontario

Ontario Ministry of Agriculture and Food
Guelph, ON N1G 4Y2
Toll-free: (888) 466-2372 (Ontario only)
Tel.: (519) 826-3100 (outside Ontario)
www.omafra.gov.on.ca/english/index.html

Quebec

Ministère de l'Agriculture des Pécheries
et de l'Alimetation du Québec
Québec, QC G1R 4X6
Tel.: (418) 380-2136
www.mapaq.gouv.qc.ca/fr/Pages/Accueil.aspx

Atlantic Provinces

New Brunswick Department of Agriculture, Aquaculture
and Fisheries
Agricultural Research Station (Experimental Farm)
Fredericton, NB E3B 5H1
Tel.: (506) 453-2666
www.gnb.ca/agriculture

Newfoundland and Labrador Department of Natural
Resources
Agrifoods Development Branch
Corner Brook, NL A2H 6J8
Tel.: (709) 637-2046
www.nr.gov.nl.ca/nr/department/contact/agrifoods/
agrifoods.html

Nova Scotia Department of Agriculture
Halifax, NS B3J 3N8
Tel.: (902) 424-4560
www.gov.ns.ca/agri

Prince Edward Island Department of Agriculture and
Forestry
Charlottetown, PE C1A 7N8
Tel.: (902) 368-4145/Toll-free: (866) PEI-FARM/734-3276
www.gov.pe.ca/agriculture

Appendix III: Chemical Symbols and Formulas

O N FERTILIZER LABELS chemical elements are referred to by their symbols rather than by their names.

Chemical Elements		
Symbol	Element	Ionic form
B	Boron	
C	Carbon	
Ca	Calcium	Ca^{2+}
Cl	Chlorine	Cl^-
Cu	Copper	Cu^{2+}
Fe	Iron	Fe^{2+}
H	Hydrogen	H^+
K	Potassium	K^+
Mg	Magnesium	Mg^{2+}
Mn	Manganese	Mn^{2+}
Mo	Molybdenum	
N	Nitrogen	
Na	Sodium	Na^{2+}
Ni	Nickel	Ni^{2+}
O	Oxygen	
P	Phosphorus	
S	Sulfur	
Zn	Zinc	Zn^{2+}

Only a few of these elements exist in nature on their own, but they are the building blocks for all the compounds that make up our universe. Many that form soluble compounds are present in ionic (electrically charged) forms when dissolved in water.

Some other common ions, formed of compounds rather than single elements:

Common Ionic Compounds

Ionic compound	Name
BO_4^{3-}	Borate
CO_3^{2-}	Carbonate
HCO_3^{-}	Bicarbonate
MoO_4^{2-}	Molybdate
NH_4^{+}	Ammonium
NO_3^{-}	Nitrate
OH^{-}	Hydroxyl
PO_4^{3-}	Phosphate
SO_4^{2-}	Sulfate

Common Chemical Compounds

Formula	Name
H_2O	Water
KCl	Potassium chloride (muriate of potash)
K_2SO_4	Potassium sulfate
NH_4NO_3	Ammonium nitrate
$(NH_4)_2SO_4$	Ammonium sulfate
$(NH_2)_2CO$	Urea
$NH_4H_2PO_4$	Monoammonium phosphate
$(NH_4)_2HPO_4$	Diammonium phosphate
$CaCO_3$	Calcite, or calcitic limestone
$Ca(NO_3)_2$	Calcium nitrate
$MgSO_4$	Magnesium sulfate (Epsom salts)

Other Compounds Listed on Fertilizer Labels

Phosphorus is expressed as the amount of plant-available phosphorus pentoxide (P_2O_5).

Potassium is expressed as the amount of soluble potash (K_2O).

Index